B.S.B.I. Conference Reports, Number Thirteen

PLANTS: WILD AND CULTIVATED

PLANTS: WILD AND CULTIVATED

A Conference on Horticulture and Field Botany
The Royal Horticultural Society and the
Botanical Society of the British Isles
2nd and 3rd September 1972

Edited by
P. S. GREEN

Published for
THE BOTANICAL SOCIETY OF THE BRITISH ISLES
by
E. W. CLASSEY, LTD.
353, HANWORTH ROAD, HAMPTON, MIDDLESEX

International Book No. ISBN O 900848 669

Printed in Great Britain at the Pendragon Press, Papworth Everard, Cambridge.

CONTENTS

PREFACE

It is a strange fact that those who are interested in native floras, especially professional botanists, often ignore cultivated plants in gardens. Conversely, gardeners tend to pay little heed to plants which are wild, unless these happen to be weeds. It was a happy suggestion, therefore, that the Botanical Society of the British Isles and the Royal Horticultural Society should hold a joint conference to examine some of the topics which are common to the interests of both Societies and their members. Perhaps, in the early stages of planning, there were some doubts as to whether such a conference could be a success, but in the event all such fears were swept away, and talk was even heard of the possibilities of another joint conference in the future. Certainly, many subjects of interest and appeal to both Societies were dealt with and the text is now published for all to read, whether they were able to attend the Conference or not.

The papers were presented in the lecture room at the Royal Horticultural Society's New Hall and during the Conference a number of associated exhibits were staged in the Hall itself. Short accounts of these exhibits are brought together as Appendix I to this Report, and although many of them were on display for the period of the Conference alone, two, which were outstanding, remained for the R.H.S. Fortnightly Show which followed. These two were the very natural-looking garden planted with native British trees, shrubs and herbs erected by the Rock Garden Department of the R.H.S. Garden at Wisley and the magnificent display of more than 160 cultivars of indigenous British trees and shrubs put on by Messrs Hillier & Sons of Winchester, a display which was all the more impressive when one realized what a limited number of native woody plants there was in the British flora.

Many of the papers were illustrated by colour slides of the plants which were mentioned and the habitats in which they grew. However, even with the increasing skills of the printer in the reproduction of colour photographs, I fear it has not been possible, in a modest report like this, to illustrate it with coloured plates. Only those who attended the Conference will have had the pleasure of seeing the many beautiful photographs which were shown.

It is customary for the Society's Conferences to be followed by a field meeting and this Conference was no exception. In order to introduce field botanists to the delights of horticulture what better place could have been visited than the R.H.S. Garden at Wisley. Conversely, within a short distance of London, what locality could be more outstanding in beauty, as well as in the diversity of wild flowers to be seen in September, than Box Hill, so well known to field botanists. Grateful thanks are expressed to Mr Peter Wanstall and Mrs Ailsa Lee, who led parties on their visit to Box Hill, and to the Director and his Staff at Wisley for their hospitality and guidance round the collections there. A joint account of the excursions to these two places forms Appendix II.

It is not usual in the Society's Conference Reports to include papers of a strictly taxonomic content. However, Dr Peter Yeo's account of *Acaena* made reference to several species the names of which were either unfamiliar or, worse still, had been employed in different ways by previous authors. It has been decided, therefore that it would be helpful, as well as appropriate, to print his formal taxonomic revision as Appendix III to this Report. As a bonus to those who like to name their plants themselves, whether from the wild or in gardens, this appendix contains an identification key to all the Acaenas with spherical heads which have been, up to now, recorded as naturalized or cultivated in Britain.

Although it did not form part of the Conference proper, on 6 September, during the Flower Show which followed, Dr S. M. Walters presented a lecture to the R.H.S., on a closely related topic very pertinent to the Conference itself. The title was "The role of botanic gardens in plant conservation" and the text, will be available for all to read, as it is to be printed in the *Journal of the Royal Horticultural Society*.

Finally, thanks are due to a number of people. Firstly to the various speakers for producing the text of their papers so promptly. The appearance of this Report without too much delay is in large part due to their help. Secondly, thanks are expressed to the Director of the Arnold Arboretum for permission to reproduce the page from *Arnoldia* which constitutes the figure on p.43 and to Mr J. E. Lousley and Messrs David & Charles Ltd for permission to reprint the figure from the *Flora of the Isles of Sicilly* which is used to illustrate Mr Lousley's paper in this Report. Thirdly, I offer my appreciation and thanks to Mr D. H. Kent for so willingly and painstakingly reading proof of the whole Report. Last, but certainly not least, I should like to state my particular indebtedness to Miss Elizabeth Young for her willing help in taking notes of the discussions which followed many of the papers and for recording the names of those who asked questions of the speakers or in other ways took part. The organizers of the Conference most helpfully arranged for the discussions to be recorded on tape and this was of inestimable help when it came to incorporating the text in this Report, but with a single microphone and occasional background noise, there would have been many gaps in this Report without the help so kindly given by Miss Young.

Royal Botanic Gardens, Kew,   P. S. GREEN
January, 1973.

HORTICULTURE AND FIELD BOTANY

held at

The Royal Horticultural New Hall, Vincent Square, London.

SATURDAY, 2nd SEPTEMBER

MORNING SESSION. *Chairman*—Professor J. Heslop Harrison.

09.45 – 10.00   WELCOME By Lord Aberconway, President
of The R.H.S.

10.00 – 10.30   "And never the twain shall meet"
(Horticulture and Botany—allies not enemies)
J. S. L. Gilmour

10.30 – 11.00   Safeguarding wild and garden plants
Dr. D. A. Ratcliffe

11.00 – 11.30   Coffee

11.30 – 12.00   How to manage garden weeds
(prepared by Dr J. G. Davison & R. J. Chancellor)
R. J. Chancellor

12.00 – 12.30   The value of herbaria for cultivated plants
J. P. M. Brenan

12.30 – 12.45   *Alchemilla*   Dr S. M. Walters

12.45 – 13.00   *Acaena*   Dr P. Yeo

13.00 – 14.15   Buffet Lunch

AFTERNOON SESSION. *Chairman*—Mr H. G. Hillier.

14.15 – 14.45   The role of nurseries   W. Ingwersen

14.45 – 15.15   The role of private gardens   G. S. Thomas

15.15 – 15.45   The role of botanic gardens   D. M. Henderson

15.45 – 16.15   Tea

16.15 – 16.45   The relevance of genetics   Dr K. Jones

16.45 – 17.00   *Hypericum*   Dr N. K. B. Robson

17.00 – 17.15   *Mesembryanthemum*   J. E. Lousley

19.00 for 19.30 CONFERENCE DINNER

Guest Speaker: Mrs Frances Perry

SUNDAY 3rd SEPTEMBER

MORNING SESSION.   *Chairman*—Dr C. E. Hubbard.

09.45 – 10.15	The principles of nomenclature	Dr W. T. Stearn
10.15 – 10.45	Problems of horticultural nomenclature	
		C. D. Brickell
10.45 – 11.15	Coffee	
11.15 – 11.45	Development of gardens plants from wild species	R. D. Meikle
11.45 – 12.15	British wild flowers in gardens	W. K. Aslet
12.15 – 12.30	Mints	Dr R. Harley
12.30 – 12.45	Snowdrops	R. D. Nutt
12.45 – 13.00	Mistletoe	Dr F. Perring

13.00 – 14.15	Buffet Lunch	

AFTERNOON SESSION.   *Chairman*—Mr W. G. MacKenzie.

14.15 – 14.45	Garden escapes and naturalized plants	
		Miss C. M. Rob
14.45 – 15.15	Literature of plants	R. Desmond
15.15 – 15.30	*Hebe*	P. S. Green
15.30 – 15.45	Arums—Lords and Ladies	Dr C. T. Prime
15.45 – 16.00	Dandelions	Dr A. J. Richards
16.00 – 16.15	SUMMING-UP By Mr D. McClintock, President of The B.S.B.I.	

MONDAY, 4th SEPTEMBER

DAY EXCURSION to visit Box Hill and the R.H.S. Garden, Wisley.

ASSOCIATED EVENTS

at the Royal Horticultural New Hall, Vincent Square

TUESDAY, 5th SEPTEMBER, R.H.S. Flower Show with special exhibit.

WEDNESDAY, 6th SEPTEMBER, R.H.S. Flower Show.

15.00	**Lecture** The role of botanic gardens in plant conservation	Dr S. M. Walters

INTRODUCTION

Lord Aberconway

President, Royal Horticultural Society

Ladies and Gentlemen, Botanists and Horticulturists.

I welcome warmly the purpose, the nature, and the form of this conference, and I am as proud as I am delighted to have been asked to open it.

We hear too much of the differences which divide botanists and horticulturists, too little of the points of common interest that they have. Where would botanists be without horticulturists? And where would horticulturists be without botanists? Each subject without the other would be infinitely more barren, botany more cold and theoretical, horticulture more aimless and superficial.

Yet the differences of viewpoint do exist, and do sometimes cause irritation. It is easier perhaps for horticulturists to grumble and to "get at" botanists than *vice versa*. I myself, I think, am held not blameless in this matter, but perhaps in my case, as some of you may appreciate, heredity is a contributory factor: the all too facile urge to pull the legs of our botanical friends is something with which I have been brought up and is often too easy to indulge in, too hard to resist.

But the more serious differences (in the sense of controversy and friction) between the two practices lie, as I see them, in two directions. First, botanists pronounce as identical, species which horticulturists know to be, as garden plants, quite distinct: the *Rhododendron* genus and the *Magnolia* genus offer numerous examples of this. Secondly we have the matter of name changes, which are frankly very irritating, though horticulturists have come to recognize the need for adherence to a logical rule and system, even if they do not readily accept the consequences of it.

Fortunately there are many botanists who are keen and gifted gardeners, and many gardeners who are experienced and knowledgeable botanists. Your two Chairmen of today, Professor Heslop-Harrison and Mr Harold Hillier, and your first, and very properly first, speaker, Mr John Gilmour, all fall within one or the other of these two definitions There are many others here today with a foot, as it were, in each camp. With their help, much bridging of gaps can be achieved at a meeting like this.

This will I am sure be a successful weekend, and it will I hope be followed by others such, at appropriate intervals. We horticulturists may have the opportunity today to impress upon our botanical friends how sad we are that, for instance, *Viburnum fragrans* is now *Viburnum farreri,* much though we love to commemorate the name of Reginald Farrer. The older ones among us may, however, be able to compliment the botanists that we may now call our old friend *Crinodendron hookeri* no longer *Tricuspidaria lanceolata* but *Crinodendron hookerianum*—almost in fact as good as old. If only the botanists would let us call that lovely plant, the deciduous *Eucryphia,* by its charming earlier name *E. pinnatifolia,* instead of *E. glutinosa,* horrid of sound, horrid of connotation, we should indeed be grateful to them. We can, however, assure them that we were delighted that they had second thoughts and removed the *"syphon"* which cast by implication a slur upon the name of dear old *Osmanthus delavayi.* If I might make a plea to both sides, could you not agree that that splendid genus named after Dr Kamel, and spelt with two 'L's' should be pronounced, as by every rule of logic it should be "Camellia", not "Cameelia"? I am always surprised and delighted when occasionally I hear someone else pronounce it as I maintain it should be pronounced "Camellia". I believe, to my regret that the Royal Horticultural Society did once decree that it should be pronounced "Cameelia", so, if the botanists could prevail upon them otherwise they would, in this matter, have my full support.

Gardeners sometimes forget that if they make a fine hybrid, and name it after their nearest and dearest, they may regret their act once the botanists have published their technical description. One keen Rhododendron hybridizer, a great friend of mine, named a splendid hybrid of his after his wife: her pleasure was great until she read a description of it, which in curtailed form ran somewhat as follows:—spotty, heavily blotched, hairy, scaly underneath. Botanists, we must remember, are only human, and the botanist in question perhaps did not like the lady very much.

I hope that tonight at our dinner horticulturists and botanists may, privately and silently perhaps, toast each other in the belief that today will have seen an improvement in the understanding of each other's point of view. If that is the feeling, this conference, which I am now happy to open, will be well worth while.

AND NEVER THE TWAIN SHALL MEET: HORTICULTURE AND BOTANY— ALLIES NOT ENEMIES

J. S. L. GILMOUR

University Botanic Garden, Cambridge

Relationships

The two halves of my title (which I did not choose myself!) appear rather contradictory, but if you consult the full text of Kipling's stanza "Oh, East is East, and West is West, and never the twain shall meet", my title would seem to imply that if a botanist and a horticulturist stand "face to face", provided they are both "strong men", they will see eye to eye on everything!

This, I think, is an over-romantic and over-optimistic view, and a more realistic way of looking at the relationship between botanists and horticulturists is as two groups of people dealing with the same material, but from rather different angles and with rather different purposes. This means that, although two such groups may profitably co-operate in many ways—there is always a risk of clashes in some of their spheres of activity. A parallel situation in everyday life would be a village hall shared by the Vicar and the local Badminton Club; they can co-operate in, say, having the windows periodically cleaned, but may both want to use the hall on Wednesday nights! On a slightly higher plane, and very comparable with botany and horticulture, is the relationship between zoologists and zoo owners; and right at the top of the scale loom the vast problems raised by competing uses for a limited area of land, as in the British Isles.

The relationship between botanists and horticulturists is, I am glad to say, nearer the friendly one between the Vicar and the Badminton Club than that between competing land-users. Their common "material" is the whole plant kingdom—but their interests and purposes are to some extent different. Horticulturists are primarily concerned in the cultivation of the plants suitable for the use and enjoyment of human beings. Botanists are primarily concerned with the pursuit of scientific knowledge about the whole plant kingdom, from every aspect. But these different purposes do not mean that there is not scope for mutual help and co-operation —though also, of course, the danger of occasional clashes.

Horticulture of help to botany

In what sort of ways can horticulturists help botanists? I will mention three. One of the most important, undoubtedly, is the skill of the horticulturist in growing living plants needed by the botanists for research purposes. This aspect is particularly evident at a University Botanic Garden such as ours at Cambridge, where, for example, we have successfully grown, among many others, tea plants (not an easy assignment!) for important physiological research—and a wide range of other examples could, of course, be quoted from the botanic gardens of the world.

Another benefit that botany has derived from horticulture is the demonstration of the enormous potential variability hidden in certain species—a variability that is swamped in the wild by crossing and competition and only becomes evident in the garden, where the variants are shielded and encouraged by the horticulturist.

My last example is in the field of conservation. This is to be dealt with fully by later speakers, and here I will only give a bare mention of the actual—and even more, potential—role of botanic and other gardens in preserving stocks of wild species that are in danger of extinction. Conservation, however, has also provided a fruitful field for conflict between botany and horticulture, as I shall point out later!

Botany of help to horticulture

Horticulture has undoubtedly helped botany. In what ways has botany helped horticulture? Again, I will mention three. The scientific study of plant ecology, linked with physiology, during the last eighty or so years has certainly helped the horticulturist to find the right conditions for the cultivation of "difficult" plants— conditions of soil, temperature, moisture and other factors—and this kind of help will continue.

A second, and quite different, sphere of help has been the writing, by botanists, of manuals specifically directed to the identification of garden plants; the names of Bailey and Rehder immediately spring to mind—but there are many other books, for example monographs of genera, written by botanists with the gardener particularly in mind.

The writing of manuals leads inevitably to the thorny problem of nomenclature. In a moment I will deal with this as a source of friction between botanists and horticulturists—but, first, it is only fair to salute the provision by botanists of an international system for the naming of plants, without which communication between horticulturists would be impossible. Further, it should be remembered that the present *International Code of Nomenclature of Cultivated Plants* originated at meetings of *botanists* at the International *Botanical* Congress held at Stockholm in 1950. Lastly, I would like to record the great help that orchid growers have had from certain botanical specialists, notably Mr Peter Hunt, in the extremely difficult and complicated business of drawing up rules for the naming and registration of man-made orchid hybrids.

Clashes in the field of conservation

So much for positive help and co-operation. Where do the clashes come? As I have hinted, they come principally in two fields where co-operation has also been fruitful—Conservation and Nomenclature.

Clashes in the field of Conservation, occur, of course, where gardeners have deliberately dug up—and thereby endangered the continued existence of—rare plants to grow in their gardens. Happily such actions are, one hopes, almost completely a thing of the past. The world-wide conservation movement of recent years has really brought home to gardeners that they must fight emotion with emotion, counteracting their desire for easy aquisition of rarities with a desire and determination to preserve the beauty and interest of the wild countryside. Other speakers will deal with the part that gardens, private and public—and especially nurserymen —can play in providing alternative sources of rare wild plants for the gardener.

Clashes in nomenclature

There is no need, at this joint meeting, to emphasize the annoyance and frustration caused not only to gardeners, but to all who use plant names (including botanists such as ecologists and plant geographers), by the frequent changing of generic names and epithets. These changes are due to three distinct causes, which can be summarized under the headings (1) misidentification, (2) taxonomic revision and (3) nomenclature.

(1) A plant may, owing to human error, be introduced under a wrong name—a name, in fact, that rightly belongs to another plant. When the mistake is discovered it must clearly be rectified—leading, of course, to a change of name. A good (or bad!) example of this was the forced exchange of names between the plants for many years called *Mahonia japonica* and *M. bealei*. Nothing can be done about this type of change, except for botanists and horticulturists to take all possible care that new introductions are correctly identified from the start.

(2) The taxonomic revision of any group of plants almost inevitably includes changes in the taxonomic status of some of the plants involved—for example, the separation of two groups of species, both hitherto regarded as belonging to the same genus, into two separate genera. Such taxonomic changes inevitably involve a change of name. For example, at one time the Redwood, and the Big-Tree were regarded as two species of a single genus *Sequoia*. Then botanists became of the opinion that they were so distinct that they should be treated as two separate genera, *Sequoia* for the Redwood, and *Sequoiadendron* for the Big-Tree.

The factors involved in such taxonomic changes are complicated and controversial, and here I would only make a plea to botanists not to propose such revisions, involving changes in names, unless they feel it absolutely essential for the progress of taxonomy and of botany as a whole.

(3) Nomenclatural changes are due to the application of the *Code of Botanical Nomenclature*—and very frequently to the

application of the "rule of priority". When the Code was first drawn up, botanists faced the problem of how to choose the "correct" name for a plant which had been given two, three, or often more, names by different botanists at different times. They decided on what was clearly the only practicable solution—namely to accept as correct the *earliest* name given in or after 1753, when Linnaeus's *Species Plantarum* was published. This criterion, however, can obviously lead to situations where a certain name has been used for many years for a particular plant and an earlier, hitherto unknown, name is then discovered, in an obscure book, for the same plant. This problem was dealt with many years ago as regards *generic* names, by including a provision in the *Code* for the *conservation* of well-known later names, and the *rejection* of (strictly correct, but unfamiliar) earlier ones. Such a provision, however, despite many attempts at Botanical Congresses, has never been applied to *specific epithets*. I will not go into the reasons for these repeated rejections —but there is no doubt in my mind that few of them are a credit to botanists. It would seem that those botanists and horticulturists not directly concerned with revisions of the *Code* must accept this position, at any rate for the present, and make the best of a very unattractive job!

In this connexion, plans are on foot for producing lists of "Recommended Names" for horticultural and other plants. The great majority of these names would be approved by botanists, but they would include a few recommendations for gardeners to use well-known, but nomenclaturally incorrect names. Such names might be regarded as "International Common Names in Latin"! Thus, if the *Viburnum fragrans* case had come up when Lists were being prepared, it could have been recommended as an "international common name" *for horticulturists,* as against *V. farreri.* The main use of such lists, however, would be to suggest which names to use, when botanists disagree over them for taxonomic reasons (e.g. *Anemone pulsatilla* versus *Pulsatilla vulgaris*). A list on the above lines is already in existence for cultivated orchids.

Well, I have said enough, I think, to show that there are ample opportunities for friendly co-operating between botanists and horti-culturists—as well as dangers of clashes and irritations to be guarded against. I will end with what I feel is one of the neatest expressions of the mutual interdependence of botany and horticulture—in the form of a couplet, defining a botanic garden, from Abel Evans' *Vertumnus* (1713)—a poem addressed to Jacob Bobart, then in charge of the Oxford Botanic Garden.

Delightful scientific Shade!
For Knowledge, as for Pleasure made

DISCUSSION

Dr A. J. RICHARDS asked Mr Gilmour whether he thought it a matter for pride that the Cambridge Botanic Garden had more British rarities in cultivation than any other place?

Mr GILMOUR replied that he presumed that what lay behind the question was the inference that they dug up British rarities? There was such a thing as growing a plant from seed; and there was such a thing as the value of having a safe, living stock of a rarity which could become extinct in the wild. He did not know whether Dr Walters would like to add anything, for he had been the inspiration behind the British collection.

Dr S. M. WALTERS said he had very mixed feelings about this but in the present situation he thought we had to face the fact that the Garden had already accumulated this kind of material and that it had very considerable conservational value. If the question was added, should the Garden have done this in the year 1960, 1940, 1930 or back into the 19th century, then one was in a very tricky area, which he hoped to deal with in more detail in his lecture to the Royal Horticultural Society after the Conference on Wednesday [See *J. Roy. Hort. Soc.* 98: in press,–Ed.]

SAFEGUARDING WILD PLANTS

D. A. RATCLIFFE

The Nature Conservancy, London

The context of conservation

The message of evolution is one of impermanence for species in time. In the continuity of life there is constant change, with one set of species eventually being replaced by another. Yet human beings are always inclined to be conservative, looking at things on the time-scale of their own life-span and cherishing and wishing to preserve that which they find interesting, attractive or useful. The gardener probably has more of an evolutionist's view of plants, from the knowledge that what is desirable can to some extent be created—by plant breeding and careful selection—and that the flexibility of nature can be manipulated in a positive way. The nature conservationist, while basically conservative in wanting to keep what already exists, is concerned with resisting the accelerated depletion of biological resources through human activity, rather than adopting a Canute-like attitude to irrevocable changes resulting from climatic influence, pestilence or evolutionary development.

Human pressures on the flora of this country consist mainly of the incidental destruction or adverse modification of habitat through various kinds of land use, as well as the deliberate collection or eradication of plants for specific purposes. Some native species, especially trees and certain grasses, are deliberately managed as an economic resource, and here the emphasis is on conservation to give sustained yield, though economic trends or fashions may dictate changes, as in the decreasing need for certain hardwood trees.

Destructive pressures

The main pressures on habitat come from agricultural intensification, commercial forestry practices, urban-industrial development, mineral exploitation, exploitation of water resources, recreational impact, "pest" control and pollution (which stems from some of the foregoing activities). Over the centuries, agriculture has contributed substantially to the richness of the British flora, by producing new habitats, communities and species which would not otherwise be here, e.g. permanent pastures and "weeds" of cultivation. The chalk grasslands, so highly regarded by botanists, have been developed through agricultural practice, but they have recently become greatly reduced in area through change in this practice. One of the most serious current problems for nature conservation is the widespread ploughing, fertilising and re-seeding of old pastures, to give new and floristically poor swards, or the straightforward conversion of these to arable land.

Crops are now "cleaned" mostly intensively by use of herbicides to eliminate weeds, and many once common species, e.g. Corn-cockle and Cornflower, have become quite rare. Hedges and hedgerow trees have been removed wholesale, and associated verges and rough corners brought into cultivation. Many field ponds have been filled in as they are no longer needed to provide water for animals. The run-off of fertilisers is also creating problems of eutrophication in rivers, lakes and bogs, with profound effects on vegetation. The draining of wetlands for cultivation has caused a very great loss of habitat and species, e.g. in the Fenland, and this effect still continues to be serious.

Commercial forestry involves such practices as clear-felling of mature woodland and replacing broad-leaved woodland by conifer plantations. The first exposes moisture and shade loving plants to excessive dehydration and frequently causes substantial loss of sensitive species, even if the woods are replanted; whilst the second usually causes a great reduction in floristic diversity because of the intensity of shade and density of needle litter which develops. The extensive planting of conifers on hill grassland, moorland and bog has caused a reduction in abundance of certain local species.

Urban-industrial development, including road making, has caused the total obliteration or severe degradation of many botanically important sites, and poses a serious threat to many which remain. Mineral exploitation has had destructive effects in many areas, and the remaining outcrops of Magnesian limestone in northern England are under threat of quarrying. The extensive removal of water-worn surface limestone for rockery stone has destroyed many limestone pavements in northern England, and depleted the populations of many important calcicoles. Exploitation of water resources has involved the flooding of valleys to create reservoirs, the raising or lowering of water levels in existing lakes, and the piping of water from streams and rivers—all with an effect on aquatic or other vegetation. Recreational pressures lead to excessive trampling on certain sites, and increased risks of flower picking and fire damage. Plants regarded as noxious weeds, whether on farmland, gardens or elsewhere, are destroyed, and there has been local elimination of species such as *Berberis* which are hosts for crop disease organisms. Pollution, from both chemical waste and sewage, has affected aquatic systems, and in combination with the run-off of fertilizers, already mentioned, has created the problem of eutrophication. In the Norfolk Broads, for instance, many interesting aquatics such as the Water Soldier have been greatly reduced and are now quite local. Atmospheric pollution has been more damaging to lowly plants such as lichens, but its effects are not fully known.

At the same time, human influence has created many new habitats, and considerably enriched the diversity of our flora, quite apart from deliberate introductions. Management of woodland has often diversified this habitat considerably, producing rides, glades and coppice structure with an abundance of species which do no flourish in closed forest. Road and railway verges are among the important plant habitats of intensively arable districts, and the great variety of "waste" habitat associated both with farms and the urban-industrial environment carries a large number of species, especially naturalized aliens. Mine and quarry spoils are often botanically interesting, and disused gravel workings usually become good sites for aquatic and marsh plants. New reservoirs increase the extent of the aquatic habitat, though their new flora usually takes longer to develop in full. In assessing threat to habitats and the needs of nature conservation it is important to remember these positive aspects of human influence.

Safeguards

Conservation of the flora in the face of adverse human pressures may be achieved by

(1) safeguarding carefully selected examples of important habitats, with their characteristic assemblages of plants, as nature reserves or some kind of protected area;

(2) promoting more general policies and attitudes relating to land management, and especially practices affecting wild plants.

Many safeguarded areas already exist. The Nature Conservancy has established 133 National Nature Reserves covering 278,000 acres and representing a very wide range of habitats and plant communities. The Conservancy has, moreover, just completed a national re-appraisal of the need for safeguarding key areas of natural and semi-natural habitats. This Nature Conservation Review, which, as the factual basis for conservation of these key areas, is intended to be published soon, has involved a nationwide survey of habitats and the selection of the most important examples. Against a background account of the range of these habitats, their flora and fauna (i.e. a statement of what there is to conserve), there is a detailed description of 711 sites judged to be of national importance, and of the criteria and methods used in assessing and selecting these. Of the 711, 397 are regarded as grade 1 and 314 as grade 2 (i.e. as slightly lower order, often duplicating the features of grade 1 sites and thus representing alternatives). The necessary measures for safeguarding these nationally important sites are being worked out in consultation with Government Departments and other interested bodies, such as the Forestry Commission, National Trust, Royal Society for the Protection of Birds, Society for the Promotion of Nature Reserves and voluntary bodies generally.

The Conservancy has in addition scheduled over 3000 Sites of Special Scientific Interest to local planning authorities and landowners, and although there are no statutory safeguards over these SSSIs, and some have been destroyed or damaged by development, this has proved to be a significant measure in protecting these habitats which are regarded as regionally or locally important (grades 3 and 4). There are also 26 Local Nature Reserves established by local authorities, and 10 Forest Nature Reserves. As well as this the voluntary bodies have between them set up a large number of reserves, some, such as the Ouse Washes in Cambridgeshire, of outstanding national importance.

Within this large number and variety of nature reserves, a high proportion of the total British flora is represented by at least one occurrence, and for certain local or rare species, the bulk of the British population is included. When reserves are established, it is, however, important that they be managed in such a way as to maintain the conditions necessary for the survival of species which might disappear or dwindle seriously if the habitat were simply left alone. Lowland grasslands are mostly artificial habitats which require a good deal of management, or they change to scrub and woodland, whilst woodland itself requires a good deal of attention over a long period. Intervention to assist propagation, as by pollination of certain orchids, can be used to boost the populations of some very rare and threatened plants. On the whole, it is more difficult to counter "natural" changes which reduce the distribution and abundance of species, especially when climatic influence is involved, or when a species is so reduced that chance factors such as disease or defoliation might eliminate a remaining colony. In a few instances, species may be disappearing through introgression of a rare plant with a more common relative to produce hybrid swarms in which the characters of the rare species are finally swamped. This appears to be happening in places to Dorset Heath, *Erica ciliaris,* and Crested Male Fern, *Dryopteris cristata.* Here, it may be necessary to isolate populations of the pure strain before the mixing has proceeded too far.

The threat from collecting

The problem of collecting focuses especially on the rare plants, and involves the taking of material for the private herbarium, garden or greenhouse, or for sale by professional gardeners and nurserymen. Collecting for the private herbarium is nowadays justifiable only for the more common and widespread plants, but it tends to lead to a desire for inclusion of rarities. Even for scientific institutions, there is nowadays little excuse for taking rare plants for herbaria, as so much material of these already exists. For some species, there are indeed more specimens stuck on sheets of paper than living individuals remaining in a wild state in Britain. The taking of rare plants for cultivation is often still more damaging, since the roots have usually to be pulled up. Many of our rare plants are common elsewhere in Europe, and there is no need to deplete the small British populations further. Nurserymen can supply acclimatized stock of many plants such as alpines, but they should not need to replenish this by native material, unless this is readily obtainable from seed.

The collecting of rare plants is mainly an educational problem. There are moves to introduce legislation to protect wild plants, but in my opinion the main value of such a measure would be in reinforcing the view that society has a moral responsibility to cherish its botanical heritage. Nothing will stop the person who really wants to collect a rare plant, except a climate of opinion in which the sense of guilt arising from such an action would become inhibitory. It is possible to deter many people from collecting showy flowers in fairly public places by means of wardening reinforced by penalties, but many of the most threatened species are not conspicuous and grow in remote or hidden places where surveillance against collecting is usually out of the question. The crucial factor is therefore the attitude of the people who go to seek these rarities. The trophy-hunting spirit dies hard: since 1957 I have personally known the populations of the very rare *Woodsia alpina* to be still further depleted in two of its best known stations. As long as this kind of unreasonable selfishness continues, conservationist-minded botanists will take the most obvious course, which is to maintain the strictest possible secrecy over a location, when this is not already well known. This is unfortunate, but inevitable. It is, however, encouraging that botanists increasingly find trophy collecting with a camera to be the most rewarding expression of the plant hunter's skill, and one which is an additional art in itself. The colour transparency is a permanent record of the living plant in its native setting, and one to enhance the delight of memories of days spent in the field, and to give pleasure to other people. Such pictures often have a considerable scientific value, besides satisfying the hunting and trophy collecting instinct.

This is not to say that the collection of rare plants for truly scientific purposes should be disallowed, but a good deal of honest heart searching is needed in making this justification. An addition to a university herbarium is not necessarily, for instance, a valid reason. If there are still institutions which require that students compile a herbarium as part of their botanical training, they must ensure that there is a complete embargo on the taking of certain rarer species—and preferably that the appearance of any of these amongst the specimens should receive a black mark! Propagation from seeds and cuttings should eliminate the need for uprooting whole specimens, even when there is a valid scientific need for living material. The problem over collecting is basically one of attitudes, and considerable progress has been made in putting over the view of wildlife as a national heritage which no one has a right to deplete further.

The deliberate eradication of species considered harmful is probably only a minor problem. Most of these "weeds" have become a problem largely through their success in spreading into suitable habitats, and they are in no danger of serious depletion. One much reduced species, however, is *Berberis,* which has been eradicated in some arable districts because it carries the serious cereal rust fungus. Paradoxically, some of the species which have appeared and flourished through the activities of Man himself are now declining and becoming rare through his current treatment of their habitat. This applies especially to the "weeds" of arable land. The poppies are still able to maintain a fair foothold on roadsides and disturbed ground generally, but cornfield weeds such as Cornflower, Corncockle and Corn Marigold have declined greatly through the "cleaning" of cereal crops. Considerable steps have been made in protecting roadside verges—an especially important botanical habitat in the lowlands—from herbicide treatment, which once threatened extensively to destroy their interest. Railway verges have so far been free from herbicidal treatment, but the present policy of selling off disused sections of railway is resulting in the destruction of some of this interesting habitat. The reclamation of waste ground and the prevailing view that all land must be put to some good purpose is a further factor militating against the survival of many showy wayside plants.

The right attitudes

Here again, there is an educational problem, in regard to current attitudes on tidiness, utility and the undesirability of "weeds". There are many convincing demonstrations of the beauty and colourful additions which wayside and wasteland assemblages of weeds make to the scene, and many people could readily accept that these are features to be cherished and not destroyed. Nevertheless, there is a deep-rooted hatred of "weeds", especially amongst gardeners, and this is fostered by the often aggressive marketing practices of pesticide salesmen. One appreciates that there are aggressive weeds of both farmland and gardens which have to be controlled, but the use of herbicides should be restricted to what is strictly necessary and not become a promiscuous onslaught on everything which could possibly be labelled "weed". There are many people, too, who cannot bear to see an unproductive square yard of land, and agricultural economics have pressed many farmers, often reluctantly, into the intensive utilization of all cultivatable land.

It is this general attrition of habitats throughout the country which is having the most serious effect on our flora. Only a more informed and benign attitude towards wild plants on the part of the people who make the changes will be likely to have much effect in reducing the rate of depletion. Positive attitudes can, however, help to enhance the botanical richness and attractiveness of the environment, where suitable opportunities exist, as in the creation of motorways, with their large verges and central reservations. Some botanists find little interest in plants which are introduced or doubtfully native, but the concept of a native species is tricky in the last analysis, and many of us are thankful for the many aliens, such as the Oxford Ragwort, which enliven another wise drab setting. It is to be hoped that the botanical garden will not become the only place where some of our cherished rare plants can be seen in life, though this medium and the seed bank are of great importance in ensuring the perpetuation of stock and in providing a supply of material both for research and for replenishing a dwindling or defunct wild population. The conservation of the British flora will depend on the successful application of all possible methods of safeguard, and not least in the kind of surveillance of declining and threatened species' populations now being undertaken by many British botanists, and co-ordinated by the Biological Records Centre of the Nature Conservancy. Autecological studies will continue to identify critical needs of sensitive species which can be applied in management, but above all, the vigilance and dedication of botanists who love their plants, and discern fresh threats before it is too late, will be the factor on which we shall probably have to depend most of all.

DISCUSSION

Mrs I. J. WEBB remarked that she had recently had the privilege of being in Western Australia, where they preserved their flora by legislation. It cost £50 to be caught with a flower in one's hand and it had been found there that this was the only way of ensuring the preservation of their flora. The population of Western Australia is relatively small and they were probably being a bit optimistic, but nevertheless, she believed that we needed to do something about such legislation in this country.

Dr RATCLIFFE replied that he did not wish to take issue on this topic and was only stating a personal view point. Legislation expressed a climate of opinion and in the long run this was what mattered. He agreed that there might be desirable effects arising from the prosecution of offenders, although he was doubtful about its true value. He was more familiar with the work of the bird protection Acts, and remarked that in Canada, people had been fined up to $2,000 for taking rare birds like gyrfalcon in the north, yet every year people still raided gyrfalcon eyries. The penalties were high but this sort of thing still continued.

Mr J. SHELDRAKE commented that with so much of garden design now in the hands of local authorities, the high cost of labour was forcing them to move increasingly towards the use of growth retardents and selective herbicides. He wondered whether a climate of opinion could be cultivated, particularly with Councillors, whereby, for example, direct labour was used in the control of weeds, rather than more and more chemicals on the ground.

Dr RATCLIFFE replied that he very much agreed. He believed that this was one of the things we had to work on, to try and wean people off the chemical control of everything where it was unnecessary. He added, however, that there was a very strong vested interest in the marketing of such products by the manufacturers and salesmen.

HOW TO MANAGE GARDEN WEEDS

R. J. CHANCELLOR and J. G. DAVISON

*Agricultural Research Council Weed Research Organization,
Begbroke Hill, Yarnton, Oxford*

Introduction

The theme of this conference is association and co-operation between Horticulturists and Botanists and we can think of few other areas where co-operation can be more useful than in the problems presented by weeds. It is only relatively recently that weeds have become a respectable scientific subject for study and although much is now known about them there still remain a number of gaps in our knowledge that could most usefully be filled. At the Weed Research Organization we place particular emphasis on finding out about the behaviour of weeds, so that control measures can be applied to the best advantage, with the greatest safety and most economically, but knowledge still lags behind need.

In this paper we give a general outline of the factors contributing to weediness in plants and how one can try to curb their often overwhelming powers of reproduction, persistence and aggressiveness. Weeds are the success stories of the plant kingdom, and it is always nice to hear about success, but when these plants interfere with one's gardening activities then we call them weeds and seek to destroy them. The definition of weed thus requires a subjective decision by the observer, for one man's weed is often another man's ornamental.

The first aspect that we consider is the number of weed seeds that occur in the soil.

Weed seeds in the soil

Freshly dug and raked soil looks neat, tidy and free of plants, yet it can harbour a quite remarkable number of weed seeds. So important is this source of weeds that it is often known as the soil seed bank, although unlike other banks the interest on the total is almost invariably compound.

A moderately dense stand of weed seedlings, and here we refer mainly to annuals, may amount to 100 individuals to the square foot. Now it has been found that seedlings that emerge, normally represent only between 2 and 10 % of the total viable seed population present in the soil. So the 100 weeds per square foot or the 4 million per acre indicate that at least a further 40 million/acre remain dormant in the soil. This figure as it happens is rather low, for many agricultural fields that are considered weedy may contain 80–100 million seeds/acre. Commercial horticulture, on the other hand, tends to generate more intensive weed problems and may have seed populations of up to 350 million viable seeds per acre. A population of this magnitude could, after a single cultivation in spring, give rise to a population of about 800 seedlings per square foot of soil. These figures give some idea of the potential size of the annual weed problem and how the weeds on the surface indicate the far greater number of seeds remaining unseen among their roots. Perennial weeds of course are rather different and we shall consider them separately later on.

The effects of cultivation

The horticulturist, the gardener and the farmer all cultivate soil to clean up residues of the last crop and any weeds that may have occurred, or to make a suitable seed bed for planting the next crop. Not only does it suit the cultivator, but the weeds too, for they have become adapted to respond by germination to the conditions of cultivated soil. Furthermore, seeds that have not lost their dormancy when buried, but have been unable to germinate at depth, will be brought up to the surface and so add to the seedling population. It has been found too that a large majority of annual weeds can emerge successfully only from a depth of 2 inches or less so that cultivation to a depth of 6 inches or more ensures that a majority of the seeds in the soil are placed at a depth from which they will not germinate. They have been placed in a sort of deposit account in the soil seed bank, although here they yield no interest, provided they are left there. This is an aspect we shall consider later with hoeing, but it is one of great importance, for deep burial of seeds generally prolongs survival even though it prevents germination.

Periodicity of germination

So far we have only considered the seed population as a whole; but the population is invariably composed of a number of different species and within these of a number of individuals of different ages and hence of differing degrees or conditions of dormancy. Each weed has its own germination periodicity too, although a majority of species will germinate mainly in spring and autumn. This may be due largely to the wide fluctuations of temperature occurring at these times of the year. A few weeds, notably the annual *Polygonum* species, germinate only in the spring, because they require a prolonged period of chilling to break their dormancy and only a short period of higher temperature in summer to reimpose it. A few species germinate mainly in the autumn, such as Parsley-piert (*Aphanes arvensis*) and Common Mouse-ear (*Cerastium holosteoides*), but the large majority of weeds germinate in both spring and autumn. There are, however, a few species that germinate only over winter, such as Common Goosegrass (*Galium aparine*) or Ivy-leaved Speedwell (*Veronica hederifolia*), or only in summer, such as Common Mallow (*Malva sylvestris*) or Black Nightshade (*Solanum nigrum*). Besides these various groups there are other weeds that are not apparently restricted by periodicity and can germinate at virtually any season of the year. Among these are the most successful invaders of garden soil, such as Common Chickweed (*Stellaria media*), Common Groundsel (*Senecio vulgaris*), Annual Meadow-grass (*Poa annua*) and Shepherd's-purse (*Capsella bursa-pastoris*). Besides this ability they can also flower and set seed throughout the year, provided the conditions are not too extreme. Unfortunately therefore one has no potential method of suppressing these ubiquitous weeds by cultivating or planting flower crops or vegetables at times outside their main germination periods, as can be done with certain other weeds.

The persistence of weed seeds

If one is good at weed control and prevents all seed production completely throughout a year, then one would expect that fewer weeds would germinate in the next season and fewer still the next. Does this in fact work in practice? It is known that seeds of some plants have a very short life-span, for example some Willows (*Salix pierotii* and *S. japonica*) have been found to lose their vitality in less than a week. On the other hand seeds of the Artic Lupin (*Lupinus arcticus*) have been thought to survive 10,000 years when under enforced dormancy in frozen soil. Weeds, on the whole, tend to favour less extreme periods, and studies have shown that the seeds of many can survive about 5–10 years in the soil. There are exceptions of course, Corn-cockle (*Agrostemma githago*) lasts just about 1 year, a feature that has contributed to its virtual disappearance in this country, Curled Dock (*Rumex crispus*) on the other hand, in one experiment that is still running, is still giving 2% germination after 80 years burial in the soil, a feature that should ensure its survival. The rate of decline of mixed weed seed populations in the soil is therefore dependant upon the species present, frequency of cultivation and, to a lesser extent, other factors such as the acidity of the soil, the water table, etc. We are fortunate that Mr Roberts at the National Vegetable Research Station has studied these problems in detail under various horticultural conditions. He has found that frequent cultivations result in a decline of the seed population of about 45% per year. In other words the population has a half-life of approximately one year. This is an interesting confirmation of the old adage that one year's seeding means seven years' weeding, for if one calculates from this, the level of a weed population after seven years would be 1·5% of the original. However, with fewer cultivations the rate of decline is less. He found that with four cultivations each year the annual drop in seed population was 32%, and with no cultivations the rate of decline was only 12% *per annum*. So without cultivations the half-life is six years and it would take about forty years to reach 1% of the original seed population. Thus the management choice,

using only mechanical control methods, is, either to cultivate
frequently, to stimulate germination, kill all the seedlings that emerge
and prevent further seeding for at least seven years (which of course
is not always possible), or to carry out a minimum of very shallow
cultivations, so as to leave the large majority of seeds at a depth
from which they cannot emerge, and maintain this system for
many years.

The control of established weeds

From the time a weed emerges it can be controlled by mechanical
means. With very small weeds any soil disturbance will probably be
effective, especially if accompanied by drying out of the soil; but
as the weeds grow older they become more resistant to haphazard
disturbance and greater precision is needed. In gardens, where the
effort available is usually measured in manpower rather than
horsepower, the hoe is the implement most widely used for weed
control. Although hoes come in various shapes and sizes the
object is always to sever the shoots from the roots just below soil
level. Some gardeners favour the Dutch hoe, which is pushed
away from the operator who walks backwards and therefore does
not leave footmarks on the newly-hoed area. Others favour the
draw hoe, which is pulled towards the operator, who usually walks
forward. Besides detracting from the appearance of the freshly
cultivated soil with this implement, there is also the possibility that
weeds that get trodden on stand a better chance of re-establishing
themselves. Many modern hoes are more versatile, having been
designed to be pushed or pulled. As the hoe is essentially a cutting
instrument the blade must be kept sharp and blades with a serrated
edge are good for cutting the wiry stems of tough weeds like
Common Knotgrass (*Polygonum aviculare*). Hoeing should be
done as shallowly as possible for two reasons. Firstly, the shallower
the cut the smaller the chance that the stem will re-root and secondly,
the more the soil is disturbed the greater the probability of further
weeds germinating.

Non-selective contact herbicides

In recent years herbicides have been developed that may be
regarded as chemical hoes. The most widely used is paraquat.
Like the hoe they are non-selective and kill almost all annual
vegetation with which they come in contact. It follows then that
they must be sprayed carefully on to the weeds while avoiding the
crop plants. They have several advantages over the hoe: they
require less physical effort; are quicker to apply, are unaffected by
soil moisture and there is no soil disturbance. Paraquat approaches
the ultimate in non-selective contact herbicides. It is effective
against most annuals. It is unaffected by rain falling shortly after
application (a boon in our climate). It acts rapidly, in summer
many weeds are dead within 24 hours. It is inactivated on contact
with the soil so there are no risks of root uptake by neighbouring
crop plants.

The rapid inactivation of paraquat by soil enables it to be used to control weeds that emerge before the crop. This technique of applying a non-selective contact herbicide just prior to crop emergence works well with slow germinators like onions and parsley. It can even be used with quick-germinating crops too if the seed-bed is prepared a while in advance of sowing the crop. This is known as the stale seed-bed technique.

Selective herbicides

It is difficult to control weeds in the crop row or even very close to it with a hoe. With non-selective herbicides it is even more difficult. There are, however, herbicides that can be sprayed over both crop and weed, that control the weed, yet leave the crop unharmed. These are known as selective herbicides and as such are a great advance over traditional mechanical methods.

In agriculture and commercial horticulture selective herbicides are used extensively in many crops; but many such herbicides are needed to deal with the wide variety of weeds encountered. This together with the need for precise application limits their usefulness in the garden, except for the control of dicotyledonous weeds in lawns.

Selective herbicides may be either contact in action (like paraquat) or translocated. The translocated herbicides, such as 2, 4-D and mecoprop, move within the plant and there is therefore less need to spray the entire weed. Their main advantage over contact herbicides is that they are more effective against perennials, which are considered below.

Pre-emergence weed control

One disadvantage of allowing weeds to become established before attempting to control them is that timing is critical. In spring and summer, when growth is rapid, the transition from an almost insignificant seedling to a well-established weed, that is both unsightly and difficult to deal with, occurs in a matter of days and some can produce seed at a very early stage. Keeping pace with weeds during this period requires constant vigilance. For many it is the predominant activity in the garden. How much better it would be if weeds were prevented from reaching the seedling stage. Frequent hoeing can achieve this, but there are situations when it is not practicable. Consider for instance trying to hoe asparagus during the cutting season or blackcurrants once the branches are weighed down with fruit.

Chemicals exist that kill weed seeds, but they require sophisticated methods of application to ensure a satisfactory effect to an adequate depth. At present they do not offer any promise to the gardener although they are used by commercial growers who would adopt them more widely if they were not so expensive. It is also possible to kill seeds by heat. The use of steam to raise the soil temperature to a level to kill weed seeds, pests and diseases has been standard practice in commercial glasshouses for many years. This method is impracticable for outdoor use except in special circumstances.

Gardeners and botanists can be distinguished by their definition of germination. To gardeners a plant has germinated when it appears above the soil surface. To the botanist of course this is emergence. Gardeners are concerned with weeds that emerge, they are not concerned about those that germinate and fail to emerge. Here we are fortunate in having a number of chemicals, the so-called soil-acting or pre-emergence herbicides, which kill weeds between germination and emergence. This method of weed control is widely used in most commercial flower, fruit and vegetable crops and there is considerable scope for using it in gardens.

The most widely used soil-acting herbicide is simazine. It will control annual weeds amongst the majority of ornamental and fruiting trees and shrubs grown in this country. It can also be used for total weed control on paths and drives and should therefore find a use in most gardens. Where crop roots occur the dose must be restricted to avoid the risk of damage, but even so a single application in the spring gives control of annual weeds for the entire season.

With herbicides such as simazine crop tolerance is based on inherent resistance to the herbicide and to depth protection. This last is the term used to describe the physical protection that results from the herbicide being confined through insolubility to the top one or two inches of soil, whereas most of the crop roots are much deeper. In addition most woody crop plants will tolerate some simazine, although the best example is maize, which, because it can grow in appreciable concentrations of simazine, has no need of depth protection.

There are other soil-acting herbicides that can be used with herbaceous plants and bulbs. Such herbicides virtually eliminate the need to control weeds in summer, together with the unsightliness of growing, dead or dying weeds, so common with other methods of control. In addition, the elimination of cultivation not only restricts further weed germination, it also provides a stable surface to walk on without leaving foot prints and without carrying loose soil about on one's boots. These herbicides are usually harmless when sprayed over vegetation, so they will not control established weeds, but equally important there is no need to avoid spraying crop plants, even if they are in full leaf.

There is also another group of soil-acting herbicides, such as dichlobenil, which not only have the residual properties of simazine but also control established annuals and kill the shoots of perennials as they emerge.

The prolonged period of control by soil-acting chemicals is due to their persistence in the soil. Understandably, concern has been expressed that the repeated use of such materials may, over the years, lead to a build-up of residues that would interfere with subsequent croppings. Residues do persist from one year to the next, but the amounts are extremely small in relation to the dose applied. There is ample experimental evidence to support commercial experience that there should be no problems, provided the manufacturers recommendations are followed.

Germination of immature seeds

If weeds manage to survive both hoe and herbicide until they reach the flowering stage, then some are capable of producing viable seeds very quickly. Scarlet Pimpernel (*Anagallis arvensis*) has given 4% germination of immature seeds fifteen days after the flowers first opened. Another garden and horticultural weed Black Nightshade (*Solanum nigrum*) has given 20% germination after the same period. Corn Spurrey (*Spergula arvensis*) has given 4% germination after as little as five days after the flowers opened. Other plants such as Common Groundsel (*Senecio vulgaris*) and Ragwort (*Senecio jacobaea*) are able to continue seed development even after the flowering shoot has been cut down and should the weather be suitable these and others, like Chickweed (*Stellaria media*), will actually re-root themselves and continue seed production. It is therefore necessary to remove these more resilient weeds after hoeing.

Seed Production

It is essential, especially with annual weeds of gardens, to control them during their initial phase of vegetative development. This phase is often of very short duration and it is remarkable how quickly many of these annual weeds can produce seeds. Annual Nettle (*Urtica urens*) for example can produce new seeds when it has only six leaves and will of course go on producing seeds over a very long period. Furthermore, several weeds will under adverse conditions of weather, especially in the autumn, or of situation, flower in a very under-developed condition. One often sees Groundsel (*Senecio vulgaris*) flowering in a crack in a wall with only one or two leaves on its stem. Groundsel, anyway, has a very short life-span and several others as well can accomplish their whole life-cycle in a mere six weeks. This, of course, results in their being able to achieve several generations during the course of a single year, so that, although they do not produce exceptional numbers of seeds, they can, by having several generations in a year, achieve the same result by a faster turn round of plants. Sir Edward Salisbury has given us the example of Groundsel, which has an average potential off-spring of 1,000 per plant. If all survived and fruited, and these in turn survived and fruited, they would amount after three generations to about 1,000 million individuals. So do not despise the lowly Groundsel as not being worthy of your attention!

Seed dispersal

Although the methods of seed dispersal are usually listed

exhaustively in botanical textbooks the relative importance of the various methods is seldom given. For the gardener or horti-culturist they are roughly divisible into three main groups. Firstly, those that are wind dispersed: the most effective of these are generally members of the *Compositae* and these have plumed seeds to make the method more effective. This group includes some of the most ubiquitous of garden weeds, for plants such as Rosebay Willowherb (*Epilobium angustifolium*), Groundsel (*Senecio vulgaris*), Perennial Sow-thistle (*Sonchus arvensis*) and Coltsfoot (*Tussilago farfara*) grow in a multitude of habitats outside the garden, in their season, and send off vast numbers of seeds, that are capable of dispersal over immense distances. These are the ones that jump the garden fence and are hardest to guard against. We have calculated that in the Oxford area about fifteen viable seeds of Perennial Sow-thistle fall on each acre of ground every year.

The second group is comprised of those that go in for active self-help. These, when ripe, violently eject their seeds away from the parent. There are a surprising number of these among garden weeds and they should never be ignored. We once had a letter from a woman gardener saying that she had been so impressed with the beauty of the Hairy Bittercress (*Cardamine hirsuta*) when she first found a few in her garden that she had transplanted some to advantageous spots. She had been delighted when they began establishing themselves, but after a year or two to her dismay she had a garden full. What, she was writing to ask, could she do about it? What indeed! The seeds of this plant are ejected forcibly up to three feet from the parent and if weeded at a mature stage the disturbance is sufficient to cause further seed ejection. The only hope is to weed it out at an early stage. It is a weed that we classify as one that you must never turn your back on. It must be weeded out immediately. Sun Spurge (*Euphorbia helioscopia*), Spring-beauty (*Montia perfoliata*), Procumbent Yellow Sorrel (*Oxalis corniculata*) and Annual Mercury (*Mercurialis annua*) are all weeds that have this unfortunate characteristic. They should conse-quently never be allowed to seed, otherwise you too will soon have a garden full.

The third group is comprised of those weeds that have no active dispersal mechanism. They tend to drop their seeds at their feet and some even, such as Nipplewort (*Lapsana communis*) and Curled Dock (*Rumex crispus*), appear reluctant even to do that. However, they must not be allowed to shed seed for this is simply storing up trouble for the future. It is known that if even limited seeding is allowed this will set back one's weed-reducing programme by eigh-teen months, while completely unrestricted seed production may set it back by at least a decade.

The shedding of mature seeds into the soil brings us back to where we started. The seed can be seen, therefore, as the vital link between one generation and the next and the reason why annual weeds are perennial nuisances. The fact that many can remain dormant in the soil for many years is probably far more important than their protracted vitality, for, if they had no dormancy at all, then exhausting the soil seed bank by frequent cultivations would be much easier. This is the reason why at the Weed Research Organization we have been testing synthetic growth-regulatory chemicals on dormant seeds to see if any will break dormancy. So far the most interesting compound is 2-chloroethylphosphonic acid, which generates ethylene. Although it stimulates the seeds of only a relatively small number of species, it does affect some of the most important weeds. Ethylene itself is already in use in America to stimulate the seeds of *Striga asiatica*. This is a serious parasitic weed that was accidentally introduced and which they are attempting to eradicate completely. This chemical has shown that the breaking of dormancy is possible and the next step now is to find further suitable weed seed stimulators.

Perennial weeds

Perennial weeds have various, completely different types of reproduction and so present different problems to annuals. Many of them, however, do produce seed, but these do not play an over-ridingly important part in reproduction.

Couch Grass (*Agropyron repens*) often produces seeds, but not necessarily in every year and they have little dormancy. Field Bindweed (*Convolvulus arvensis*) is also spasmodic in its seed production, being mostly produced in hot summers, but it does have dormancy, which is due to hardseededness and can only be overcome artificially by treating them for half-an-hour with concentrated sulphuric acid. Other weeds have impediments to seed production. For example, Hedge Bindweed (*Calystegia sepium*) and Creeping Yellowcress (*Rorippa sylvestris*) appear to be self-sterile, while the Stinging Nettle (*Urtica dioica*) and Creeping Thistle (*Cirsium arvense*) are dioecious. The latter only produces seed if the two sexes are less than 100 feet apart and this apparently does not often occur, for seed is relatively uncommon. The thistle down that blows all over the place in summer is of no significance whatever, because this plant when it does produce seed separates pappus and seed at an early stage, so that the seed is normally left behind in the head. Coltsfoot (*Tussilago farfara*) produces seed, but it only remains viable for three months. Yet others just do not produce any seed at all e.g. Slender Speedwell (*Veronica filiformis*), Blue Sow-thistle (*Cicerbita macrophylla*), Pink-flowered Oxalis (*Oxalis latifolia*) and Winter Heliotrope (*Petasites fragrans*).

Vegetative reproduction

Despite their deficiencies in seed production many of these perennial weeds have developed very effective methods of vegetative reproduction. The parts that are regenerative vary from species to species. Some have creeping stems that are below ground, as in Japanese Knotweed (*Polygonum cuspidatum*) and Ground Elder

(*Aegopodium podagraria*) or above ground as in Creeping Bent (*Agrostis stolonifera*) or Creeping Buttercup (*Ranunculus repens*). Some species have creeping roots, such as Perennial Sow-thistle (*Sonchus arvensis*) and Field Bindweed (*Convolvulus arvensis*). While others (often biennials) have swollen, non-creeping tap roots, for example Dandelion (*Taraxacum officinale*) and Spear Thistle (*Cirsium vulgare*). All weeds with creeping parts have very considerable regenerative ability, indeed even very small fragments are capable of regeneration so that if they are to be dug out every piece must be removed or the digging repeated until this is accomplished. On the other hand tap-rooted species such as Hemlock (*Conium maculatum*), Hogweed (*Heracleum sphondylium*) and Curled Dock (*Rumex crispus*) if spudded out to a depth of three to four inches will not regenerate again as the lower levels of the root are incapable of producing new shoots. The exception to this rule is the Dandelion (*Taraxacum officinale*), which can regenerate from all levels. However, it is susceptible to growth-regulator weedkillers, so there is really no difficulty.

A few weeds produce bulbils both above and below ground and this group includes some of the most intractable problems in horticulture and gardens. Bermuda Buttercup (*Oxalis pes-caprae*) is a very serious problem to horticulture in the Isles of Scilly, while the two pink-flowered relatives (*Oxalis latifolia* and *O. corymbosa*) are frequent in gardens and nurseries in south-western or southern England. Another weed with bulbils, the tetraploid Lesser Celandine (*Ranunculus ficaria* subsp. *bulbifer*) is frequently a problem in gardens. The bulbils of these species can remain dormant in the soil for prolonged periods and at present there are no effective means of control against *Oxalis* and certainly no means of breaking the dormancy of bulbils. Young pigs are said to root for and eat the bulbils of *Oxalis*, but who is prepared to release piglets into their garden!

How to control perennial weeds by cultivation

Perennial weeds are much more difficult to control than annuals and although any weed will eventually succumb to repeated destruction of its aerial parts, the frequency and duration necessary is formidable. In general, therefore, a more positive approach is needed and there are two ways of doing this. Firstly by elimination of the regenerative buds, without which regrowth is impossible, and secondly by depletion or exhaustion of the food reserves.

Species vary in their regenerative ability. In Couch Grass (*Agropyron repens*), once all the buds on the rhizomes have grown out or been destroyed no further growth will take place, regardless of the food reserves remaining, because this plant is incapable of producing adventitious buds. In a small garden, such perennials with creeping rhizomes, especially ones which never grow deep like Couch, are best dug up. However, this is all too often complicated by the rhizomes growing under a lawn or among treasured plants which would suffer if disturbed. In contrast, perennials with creeping roots such as Field Bindweed (*Convolvulus arvensis*) appear capable of producing adventitious buds at any position along the entire length of their roots, and are much less quickly exhausted. Usually they also possess vertical roots that can grow to remarkable depths.

In larger areas repeated cultivations that break up the underground parts into fragments, and so stimulate buds to grow, is effective with time or repeated applications of a contact herbicide can be a suitable alternative.

Control of perennials by chemicals

One of the most important havens for perennial weeds in gardens is, of course, the lawn. Here may be found many species such as Plantains (*Plantago* spp.), the Daisy (*Bellis perennis*), Yarrow (*Achillea millefolium*) and Clover (*Trifolium repens*), and several others that are a problem nowhere else in the garden. Hand-weeding a lawn is extremely tedious and time-consuming, frequently resulting in unsightly bare patches. This is the one area of the garden for which selective post-emergence herbidices are well suited. By choosing an appropriate growth-regulator herbicide, such as 2, 4-D or mecoprop or mixtures of them, the dicotyledonous perennial weeds can be controlled without harming the lawn.

The same herbicides can be used elsewhere providing care is taken to avoid "treating" desirable plants. Herbicides are effective against a variety of dicotyledonous perennial weeds. Ideally, application should be made when growth is vigorous, just before flowering. It takes two, three or even four weeks for the shoots to die, but the chemical moves into the underground parts and prevents regrowth in the year of treatment and restricts it in the following year, which is a lot more than can be achieved with a single hoeing or cultivation. This explains the willingness of many gardeners to spend a considerable amount of time applying weedkiller to individual leaves and shoots with a brush or even in dipping shoots into a solution of herbicide. The task should be much easier now that special formulations in jelly are available. There is also the 'herbicide glove', which allows individual shoots to be stroked with a pad impregnated with weedkiller. The herbicides just mentioned are only effective against dicotyledonous plants; but there are other herbicides which are equally effective against grass weeds.

Soil-acting herbicides

That prevention is better than cure, is true of all weeds, especially perennials. When considering annuals, we mentioned that dichlobenil, a soil-acting herbicide, also controlled emerged seedlings. In addition it controls many perennial weeds. A single application in the spring will give control for several months. Weeds such as Ground Elder (*Aegopodium podagraria*), Creeping Thistle (*Cirsium arvense*), Couch Grass (*Agropyron repens*), Horsetail (*Equisetum arvense*), Stinging Nettle (*Urtica dioica*) and Coltsfoot (*Tussilago farfara*) are all controlled. It can be used in bush and tree fruits and among many ornamental shrubs and trees, as well as on unplanted areas. Unfortunately it cannot be used among herbaceous plants. Normally, application should be in winter or early spring, but can be later, especially when localized for the control of individual weeds or clumps of plants such as Docks (*Rumex* spp.) and Stinging Nettle (*Urtica dioica*). Unlike most weedkillers it is not applied as a spray. It is sold as granules and applied as such by sprinkling them on the soil. This makes it particularly convenient for treating both isolated clumps and in among shrubs where spraying would be difficult.

Conclusion

In conclusion, we would emphasize that to manage garden weeds successfully it is essential to know your weed. Firstly, by correct identification and secondly, by getting some knowledge of its characteristics and behaviour, for some weeds can behave in a most unscrupulous manner. Then, armed with hoe or herbicide, prevent them from multiplying and with time the problem will decline.

DISCUSSION

Mrs B. H. S. RUSSELL remarked that in her garden she had practically all the weeds that had been illustrated except the Blue Sow-thistle, but worst of all were Bindweed and Horsetail. What could be done for them?

Dr DAVISON asked where they were growing in the garden or, in other words, what were the crops that were growing near the Bindweed and Horsetail?

Mrs RUSSELL replied that they were growing practically everywhere.

Dr DAVISON replied that the *Convolvulus* (or *Calystegia*) was relatively easy to control as long as one of the lawn-type weed-killers was applied; in other words, something containing 2,4-D, provided, of course, that it was kept off any desirable plants in the garden, which was one of the difficulties. Many people overcame this by laboriously going round and painting individual leaves with a paint brush, with a measure of success. If one had less time one should try to spray the weeds carefully.

He further commented that the important point here was that if one was spraying with weed-killers one must realize that the concept was completely different from when one was spraying insecticides and fungicides. With these one normally tried to get a uniform cover on everything and one was not worried if one got a little bit of drift, so one tended to use a fine spray which was rather ill-defined as to where it went. But with weed-killer it was very important that one had a low pressure spray which was directed most carefully. Ideally one should use a nozzle that would give a coarse spray. A colleague at work had recently developed a glove which received weed-killer from a pack on one's back; this at present was only available to farmers but eventually it would be a boon to private gardeners who had only the odd weed they wished to get rid of.

He went on to remark that he had not forgotten the Horsetail! Where one had this growing amongst perennial woody plants, be they shrubs or fruit trees (but certainly not herbaceous plants, which one was unlikely to be able to save), one could certainly use the chemical called dichlobenil, which could be bought as Casoron G. If one put this on in the spring it would control most of one's weeds for the rest of the season.

Mr R. W. JASPER remarked that he was surprised there had been no mention of smother-mulching, using either black polythene or vegetable debris such as grass mowings, or indeed ground-cover plants. He asked whether an elaboration could be made on this point.

Mr CHANCELLOR replied that many of the weeds one was concerned with in gardens seemed to be able to grow through these. He agreed that black polythene was effective, but wondered who wanted to put down a lot of black polythene in their garden! He said that they had received a number of suggestions for mulches from correspondents, suggesting anything from pine-needles to newspaper, but on the whole they had tended not to be effective.

Mrs P.M. GALBRAITH asked what one did with ones grass-clippings if one used a selective weed-killer on the lawn. Could they be used on the flower beds?

Dr DAVISON replied that the first answer was that, if one decided to use any kind of weed-killer, then one should read the label before it was applied. He thought that one would find that what one should do was set out on most labels. He felt that this question gave him the opportunity to plug the Ministry of Agriculture's Approval Scheme. When one was buying a packet of pesticides one should look for a small "A" with a crown on top, which meant that the Ministry had approved the manufacturer's recommendations. As far as the lawn weed-killers were concerned, he thought that the instructions with the majority of brands said that grass from the first one or two cuts after application should not be used for mulching immediately. It should in fact be composted and after a season or so, when one would normally be using the compost anyway, there should be no problem. The main risk from using this type of weed-killer in the garden was not what was done with the grass mowings, but what was done with the watering can or spraying machine after it had been used to apply the weed-killer.

THE VALUE OF HERBARIA FOR CULTIVATED PLANTS

J. P. M. BRENAN

Royal Botanic Gardens, Kew

The classification of cultivated plants

The theme of this Conference is, I quote, "association and co-operation between horticulturists and botanists". I hope that holding it in the Royal Horticultural Society Hall does not mean that the botanists are playing an "away match".

I do not want to stress the distinction between sheep and goats before an audience like this—naturally I mean that your interest is in plants rather than animals!—but the study of cultivated plants is not to the taste of every taxonomic botanist. This is putting it gently; in fact I have seen botanists quite enraged with the odd leaf from that exciting plant in a pot in the front room sent in an envelope for identification. To say that this shows the sender's boundless faith in the botanist's ability to perform miracles does not help the situation!

Gardening may share with watching football a claim to be the national pastime, so it seems somewhat of a paradox that we are sometimes reluctant to study too closely what grows in gardens. To name our cultivated plants we must still too often rely on the works of Bailey or Rehder, admirable in content, but transatlantic in origin and scope.

To do the botanists justice, it must be said that the state of classification in many cultivated genera is chaotically difficult. The botanist may consider that the species in some cultivated groups —succulents are an obvious example—have been devalued more than is permissible, even in these times of inflation elsewhere. A natural tendency to promiscuousness in plants is often aided and abetted by the horticulturist; the results may be wonderful, but they are hard to fit into any rational scheme of classification. Tolerance and understanding of needs and problems seem to be the key.

The importance of herbaria

The interests of horticulturists and taxonomic botanists find common ground in the classification of cultivated plants. Herbaria have an important part to play in this; but how important? This is the first of two questions I shall discuss now. The second question, which is linked to and partly overlaps the first, is: what value have herbarium specimens in the study and recording of cultivars?

In discussing the first question I shall exclude for the time being the cultivars, because they present their own peculiar difficulties which will be explained in dealing with the second question.

Herbaria, which are normally collections of dried and pressed plant specimens systematically arranged for reference, come in many shapes and sizes, general or special, but not many are primarily for cultivated plants. The select few include some famous names, however, the Arnold Arboretum and the L. H. Bailey Hortorium, for example, which will be mentioned again later on. At both of these the scientific study of cultivated plants is carried out, and both are linked with celebrated universities, Harvard and Cornell respectively (Kobuski, 1958; Sutton, 1965).

These are special cases, perhaps, and let us now look at the significance of cultivated plant specimens in the more general herbarium. I wish to make the case that good specimens of both wild and cultivated plants are both valuable in their own right and mutually beneficial.

Herbarium specimens of cultivated plants provide a record of taxa which are or have been in cultivation; information about origins and dates of introduction, as well as differing variants of species; and also their success (or fate) subsequently. By piecing together the evidence from a series of specimens it may be possible to assess the climatic tolerance of a cultivated plant, and even to plot it cartographically.

Evidence about their genetic history may also be given, for instance, where 'and when mutants arose. To take one example, *Mimosa invisa* is a native of South America, strongly spiny in the wild, and cultivated as a ground cover and green manure in the tropics. In the early 1950's an unarmed variety appeared in Indonesia, which was described by Adelbert (in *Reinwardtia* 2: 359 (1953)) as var. *inermis*. Herbarium specimens provide a record of the spread of this variant to Africa, mainly it seems from one experimental station to another. Similarly, evidence may be provided by herbarium specimens of accidental hybridization between a cultivated species and others.

Cultivated material in the herbarium is often valuable to compare with and add to the information about the same taxa in the wild. Particularly where a botanic garden is associated with the herbarium, plant specimens grown in the former may be fuller, better preserved, better annotated and may show stages in development otherwise unrepresented. Also it may be easier to supplement them with "background material"—photographs, drawings, material in liquid preservative, etc. In the same way, but conversely, the taxonomy of cultivated plants may be thus better understood. It is the policy of the Bailey Hortorium to supplement herbarium specimens of cultivated plants whenever possible by others collected in the wild. In this way changes and modifications which have occurred as a result of or during cultivation may be observed and explained. The question of where the specimens of cultivated plants should actually be placed in the herbarium has given rise to discussion and argument. At Kew the practice is for the cultivated specimens to be placed at the end of the genus in their own distinctively labelled genus-covers.

Finally, but by no means least important, herbarium specimens can serve as a record of experimental work in its widest sense— breeding programmes, cytological observations, biochemical analyses, etc—carried out on cultivated plants. It still needs emphasizing that these specimens may provide the only means of verifying the identity of plants used in past experimental work. It is sad to reflect how often rigorous precautions and techniques in carrying out experiments have been allied to a sadly naïve belief in the infallibility of names on seed packets and on labels in botanic gardens!

There is thus, I submit, a strong case that herbaria should preserve specimens of both cultivated plants and wild plants transferred to cultivation. This has been carried out for many years in the Kew Herbarium with benefit to botanical and horticultural studies.

The value of herbaria in the study of cultivars

I now come to my second main topic, the value of herbaria in the study and recording of cultivars. What, firstly, is meant by this inelegant portmanteau word, cultivar? The *International Code of Nomenclature of Cultivated Plants* (1969) defines it in Article 10 as follows:—"an assemblage of cultivated plants which is clearly distinguished by any characters (morphological, physiological, cytological, chemical, or others) and which, when reproduced (sexually or asexually), retains its distinguishing features". In Article 11 we learn that cultivars are of various kinds, lines, clones or genetically distinct assemblages of individuals, and it is clear from Article 12 that cultivars may be recognized within hybrids as well as species.

Some additional comments on all this may be helpful. When a cultivar is a genetic variant within a species it may directly correspond with a genotype which occurs or is able to occur in the wild. But its existence may there be transient. When a cultivar is a hybrid segregate, it may sometimes correspond with something occurring naturally, but more often not. Examples of the latter are crosses between species which could not possibly meet in the wild, or secondary crosses involving hybrids, themselves of cultivated origin.

The nature of the differences characterizing cultivars needs discussion, for it can greatly effect their recording and documentation. The recognition and differentiation of cultivars arise mainly from the practical needs of horticulture and agriculture. In comparison with species and orthodox varieties the differences between cultivars may be few and slight. A change of shade in the colour of the petals may alone differentiate one rose cultivar from another. Two closely related but distinct cultivars of wheat are separated solely by a difference in shape of the scutellum. Many of the differences are matters of colour or morphology of parts, for example the copper beech, the simple-leaved ash, the cut-leaved alder etc. Some cultivars, however, particularly in trees, differ in habit, for example the weeping or dwarf habit of many conifers. These are features to be verified by simple observation.

Other differences may be much more subtle, and these may be termed "behavioural differences". They are found mainly, perhaps, though not exclusively, in crop plants, when physiological differences in behaviour may be commercially important in, say, enlarging the range of climatic tolerance in a crop, or in ensuring an earlier harvest. Such differences can not be assessed from herbarium specimens alone, but only from the performance of the living plant, preferably when grown as a crop.

The registration of cultivar names

Although the nomenclature of cultivars is controlled by the International Code, the actual application of cultivar names is much less firmly governed. This is in strong contrast with species and varieties, the application of whose names is rigidly controlled (some might say too rigidly!) by the identity of type-specimens. A cultivar must be described, but there is no obligatory linkage with a preserved specimen. For species and varieties herbarium specimens are legally necessary, but not for cultivars. Recommendation 39C states that "when appropriate a preserved specimen and/or illustration........should be deposited in a public herbarium and be cited in the description". This can hardly be retrospective, however.

Under the provisions of the International Code the recording of cultivar names is carried out by various Registration Authorities, comprising societies, botanic gardens, research institutes, etc. in a number of different countries. Sometimes registration may be not only a matter of following the International Code, but a statutory obligation. For example, in Britain the Plant Variety and Seeds Act of 1964 gave protection to breeders by introducing a series of schemes for plant-breeders' rights, and the registration of names of cultivars of certain genera and species is legally compulsory in Britain. Each Registration Authority under the International Code is responsible for cultivars in one or more plant groups. For example, the Royal Horticultural Society does *Lilium, Narcissus,* etc., the American Begonia Society does *Begonia,* the National Institute of Agricultural Botany in Cambridge covers agricultural and field horticultural crops, (except where the crops are covered by the 1964 Act when it is the responsibility of the Plant Variety Rights Office in London), the Arnold Arboretum various woody genera including *Chaenomeles* and *Forsythia.*

The actual procedure of Registration Authorities varies, and as examples I wish to look a little more fully at three of them, the Arnold Arboretum, the National Institute of Agricultural Botany at Cambridge, and the Plant Variety Rights Office in London.

The Arnold Arboretum has published an enumeration of cultivars in the genus *Chaenomeles* (Weber, 1963), amongst others. The paper, after a general account of the horticultural history of the genus, contains a straight alphabetical list of all known cultivar names in the genus. The names are divided into four categories: synonyms, valid names of cultivars which are currently grown and available in the U.S.A., extinct cultivars, and doubtfuls. This is followed by a descriptive list of cultivars arranged alphabetically under their species and hybrid groups. A page from this list is illustrated to show the method of treatment (Fig. 1). It should be noted that the descriptions are brief, rarely more than one or two lines, often with little more than a mention of the flower-colour. There are no illustrations. There are other lists of comparable standard, for example *Weigela* (Howard, 1965), *Lantana* (Howard, 1969) and *Ulmus* (Green, 1964). The standard form sent to those wishing to register new cultivars at the Arboretum asks, among other things, for a herbarium specimen. I am grateful to Dr Gordon P. De Wolf and Mr Robert S. Hebb, both on the staff of the Arnold Arboretum, for much generous help and information.

The National Institute of Agricultural Botany, Cambridge, whose Director Dr Peter Wellington, has given me much help, grows cultivars of a wide range of crops and vegetables, assessing them for distinctness, stability and uniformity, and testing them under cultivation, as a preliminary to advising the Ministry of Agriculture, Fisheries and Food on the official listing of cultivars and giving advice on the choice of variety to grow in a particular environment (see Kelly, 1968). The names of new cultivars can be registered under the Code, or in appropriate cases advice is given by the Plant Variety Rights Office. It is desirable to distinguish here between testing for distinctness and assessment of performance for crop production. The former is not directly related to choice of variety, but ensures the elimination of synonyms and homonyms and provides a definitive seed sample, comparable to a botanical type-specimen, which defines the new cultivar. The latter, on the other hand, enables the agronomic performance of two distinct cultivars to be assessed and compared so that A can be recommended as better than B for a specified purpose and under defined conditions. Cultivars are described on elaborate summary and recording sheets. That for peas contains 87 characters, many of them subdivided into several categories; that for lettuces 64, and so on. But, and this is important, the written schedules are supplementary to and to be taken in conjunction with comparative plots established from viable seed stocks. These definitive stocks are renewed as needed from the original breeders, and when this is no longer possible, consideration is given to whether or not another commercial source is available; otherwise it is regarded as obsolete. The cultivar may be retained by the Institute for academic interest or as part of a reference collection. The point of reference here is viable seed plus a written description.

Selection of W. B. Clarke, San Jose, California, probably no. 330, sent to Kluis Nursery, Boskoop, Netherlands, around 1946.

'Maulei' (*Pyrus maulei* Mast. Gard. Chron. II. **1**: 756, *f. 159*. 1874). Flowers salmon-pink to orange, single. Named for Messrs. Maule, nurserymen at Bristol, England, who introduced it from Japan in 1869. The strain of **C. japonica** introduced by the Maules from Japanese gardens differs from the alpine strain introduced by Sargent in growing slightly taller and in having a heavier fruit production.

'Maulei Seedlings' (Slocock Nurs., Woking, Engl., Cat. 1958–59). Flowers orange-flame. Probably not a clone, but only selected seedlings of **C. japonica** 'MAULEI'.

'Mawlei' (*C. japonica Mawlei* Buyssens Nurs., Uccle, Belg., Cat. 1933, without description) = 'MAULEI'.

'Moulei' (Van Geert Nurs., Anvers, Belg., Cat. 1896, without description) = 'MAULEI'.

'Nana' (cult. at the Univ. of Connecticut, Storrs, Conn.) = 'PIGMANI'.

'ORANGE BEAUTY' (Jaarb. Boskoop 1954: 116. 1954, without description). Flowers orange, single; Dutch selection, before 1954.

'Pigmaea' (Clarke Nurs., San Jose, Calif., Wholesale Price List Nov. 15, 1935) = 'SARGENTII'.

'Pigmaea' (*C. lagenaria* Pigmaea, Light Tree Nurs., Richland, Mich., Price List 1958) = 'PIGMANI'.

'PIGMANI' (Anonymous, Pl. Buyer's Guide 93. 1958, without description). Flowers red-orange, single, often unisexual. Selected in Kallay Nursery, Painesville, Ohio, in 1954, under the name of 'Pigmaea'.

'PLENA' (*C. maulei* f. *plena* Iwata, Jour. Agr. Sci. [Setagaya], **5**(4): 36. 1960). Flowers double; flower color and origin unknown, before 1960. In Japanese gardens.

Var. **pygmaea** (*C. japonica* var. γ *pygmaea* Maxim. Bull. Acad. Sci. Petersb. **19**: 168. 1873). Branches often subterranean. The type specimen of this variety was collected around Yokohama, Japan. The name should not be applied to a cultivar. It is not a synonym of **C. japonica** var. **alpina** Maxim.

'Pygmaea' (*C. japonica pygmaea* Chenault Nurs., Orléans, Fr., Cat. 1910–11) = 'SARGENTII'.

'Pygmaea alba' (cult. at the Stanley M. Rowe Arb., Cincinnati, Ohio). This name is not acceptable according to the International Code of Nomenclature for Cultivated Plants which prohibits new names of cultivars in a Latin form. We propose to name it 'DOROTHY ROWE'.

'Pygmy' (Linn County Nurs., Center Point, Iowa, Cat. 1960) = 'SARGENTII'.

'Sargentiana' (cult. at the Wageningen Arb., Wageningen, Neth.) = 'SARGENTII'.

'SARGENTII' (*Cydonia sargenti* Lemoine Nurs., Nancy, Fr., Cat. no. **143**: ix. 1899). Shrub more dwarf than the typical form of the species; flowers

[29]

Fig. 1 A page from the list of cultivars in the genus *Chaenomeles* by Dr Claude Weber in *Arnoldia* 23 (1963). Synonyms are indicated by italics; maintained cultivar names are indicated by capitals; varieties and hybrid groups are in bold-face type. Reproduced by kind permission of the Director of the Arnold Arboretum.

The National Institute of Agricultural Botany maintains considerable collections of dried specimens, though not pressed in the normal herbarium sense, for crops such as cereals, where they are very useful in identification. For each cultivar, in fact, sets of cards are prepared illustrating the range of characters shown by the grains, glumes, etc. In peas, where group classification of cultivars is based on seed and pod characters, dried specimens and photographs are kept for all varieties in the Institute's Common Knowledge Collection. The Institute finds that herbarium material is particularly useful when distinctness depends on morphological features, but not when physiological differences such as maturity are more important.

The Plant Variety Rights Office is not a Registration Authority under the Code, but has statutory functions of a similar nature under the 1964 Act. The Office is mainly concerned (as far as decorative plants go) with *Rosa, Chrysanthemum, Dianthus, Freesia* and *Rhododendron,* and cultivars of these genera are recorded on written forms similar to those used by N.I.A.B. The form for roses is a technical report form internationally agreed under the Union for the Protection of New Varieties of Plants (U.P.O.V.), and covers 68 characters, each often subdivided. My thanks are due to Mr L. J. Smith, Controller of the P.V.R.O., and to Mr J. M. Evans, also on the staff, for information about their procedure.

The Plant Variety Rights Office has concluded that pressed herbarium specimens of protected cultivars would be of very limited value in relation to their existing system of recording, though the possibility has been considered since the Plant Variety and Seeds Act of 1964 came into effect.

The adequacy of the present system

We have seen the directions taken by current practice in cultivar registration. They are not uniform, but the differences probably reflect varying needs in different groups of plants. There is no doubt that the registration system is essential, particularly where there are statutory requirements. Is the existing system, which in general functions reasonably well, sufficient? Can it be left much as it is? Are alterations or additions needed, and if so has the herbarium specimen any role here?

These questions have more than a theoretical academic interest: during the past year the question has arisen at Kew of how does one verify whether a cultivar is correctly named or not? What reference or standard should one use? It is not easy to give a simple answer. With cultivated crops the point of reference is viable seed supplemented by a very full description. The Royal Horticultural Society have kept a record of certain orchid cultivars by means of paintings, but with other genera it may be just a description varying from a detailed schedule to no more than a line or two.

In answering the question of whether the existing system is adequate, it is important to bear in mind the size of the job. To take one example, over 15,000 cultivar names are registered for African Violets alone. Any elaborate system will consume manpower, money and time out of proportion to the importance of the task. No herbarium curator, for example, will be enthusiastic about the prospect of having 15,000 voucher specimens of African Violets. In addition what is the scientific or horticultural value of this mass of names? Many, perhaps most, represent genotypes which rapidly become obsolete or even vanish altogether, ousted by new cultivars, or new names! The conclusion is inevitable that any modifications of the present system, which probably represents the basic essentials, must be on a very rigorously selective and partial basis.

The role of herbarium specimens in registration

There can be little doubt, however, that herbarium specimens of cultivars have some part to play. For cultivars defined mainly by behavioural or physiological characters, they will be of no value, but where the differences are mainly morphological then they may be useful. But how can this part be defined and limited? We have seen that herbarium specimens, when appropriate, are a recommendation for *new* cultivars, but what about all the previously existing ones?

1. Firstly, the cultivars of herbaceous plants propagated by seed are usually quickly changing and of little long-term importance. Herbarium specimens are likely to have little significance here, even as a record of morphologically different cultivars. Only occasional exceptions to this are to be expected, for example where a special breeding programme is involved.

2. Secondly, the recording of clonally propagated, morphologically distinct, cultivars of herbaceous plants may be usefully supplemented by herbarium specimens. This can only be done selectively and an important criterion must be long-term persistence in commercial horticulture. Some currently available cultivars in the gesneriad genus *Achimenes,* which is propagated by rhizomes, go back to 1855. No doubt similar histories could be found elsewhere, for example among Daffodils and Narcissi.

3. Thirdly, the greatest scope for the use of herbarium specimens of cultivars is likely to be among shrubs and trees where even one generation may last for many years. Specimens are especially valuable where leaf-shape is significant. Again, long-term persistence of a cultivar in commercial horticulture must be an important factor.

It is, I think, important that herbarium specimens of cultivars should normally not have a legal status. They should not be types, but just vouchers, points of reference, but authenticated as much as possible.

A programme of work having been proposed, it is of course an old-established and valuable technique to delegate it to someone else. And this of course raises the question of who is to select and preserve such herbarium specimens. I am sure that I speak for all curators here in saying that no one institution will be prepared to carry the whole can—certainly not Kew! Some herbaria, in their capacity as Registration Authorities, already file herbarium specimens of their appropriate cultivar groups. Perhaps this points the way that the appropriate Registration Authorities should advise and if need be act, either by themselves or in association with a local herbarium or some other one where research on the taxonomy of the group concerned has been carried out. This action might involve the establishment of limited reference collections, preferably on the basis of one genus or group in one place, of herbarium specimens of selected cultivars for reference purposes.

I hope that I am right in thinking that I have focussed on a problem that is in need of thought and discussion.

During the writing of this paper I have received much help from many people. Some I have already acknowledged, but I would like to mention particularly Mr Peter S. Green, Deputy Keeper of the Herbarium, Royal Botanic Gardens, Kew, and Professor Harold E. Moore Jr. of the L.H. Bailey Hortorium, Cornell University.

REFERENCES

GREEN, P. S. (1964). Registration of cultivar names in *Ulmus*. *Arnoldia* **24**: 41–80.
HOWARD, R. A. (1965). A check-list of cultivar names in *Weigela*. *Arnoldia* **25**: 49–69.
——— (1969). A check-list of cultivar names used in the genus *Lantana*. *Arnoldia* **29**: 73–109.
KELLY, A. F. (1968). The work of Systematic Botany Branch. *J. Nat. Inst. Agric. Bot.* **11**: 246–252.
KOBUSKI, C. E. (1958). The horticultural herbarium. *Arnoldia* **18**: 25–28.
SUTTON, S. B. (1965). The herbarium introduced. *Arnoldia* **25**: 37–40.
WEBER, C. (1963). Cultivars in the genus *Chaenomeles*. *Arnoldia* **23**: 17–75.

DISCUSSION

Mr D. McCLINTOCK commented that one of the things he found most often missing from notes on herbarium specimens was any indication of flower or leaf colour. In cultivars it mattered a great deal, for in one cultivar they might be pink and in another red, yet they both dried the same fawny colour. He felt it very important that some standard colour chart should be used to record the colours of the fresh flowers and that this information ought to be recorded.

Mr BRENAN agreed that this kind of information was more important for cultivars, where the genetic basis was often very limited, than for species, where there might be a wide range of natural variation in such matters as colour. If the colour of one individual in a wild population was described too closely one might exclude a large part of the spectrum of natural variation. But in cultivars a record of colour was obviously valuable.

Dr W. T. STEARN said he would like to illustrate the great historical and genetical value of proper specimens of cultivated plants. In the year 1930 he was monographing the genus *Epimedium*. There were a lot of names in the literature, including names for cultivars. Very fortunately the firm of Vilmorin in Paris had made a herbarium. It was the only firm that ever did this in the middle of the last century and in the herbarium he found all the Epimediums that had been in cultivation, and under their cultivar names. He was able to look at their pollen, which revealed that the cultivars were of hybrid origin; so he was able to sort out their taxonomy, and later found that the cultivars had persisted in cultivation. He considered that this was a very good example of the great value in the stabilization of names, and in genetical research, of cultivated specimens placed in a herbarium.

He further remarked that there was one celebrated herbarium in the United States, without mentioning names, in which there were a lot of specimens of cultivated plants, and a certain curator put them one after another into the waste-paper basket. Very fortunately, a student, at considerable risk to his Ph.D., came along afterwards and rescued them, and they are now in the Bailey Hortorium.

ALCHEMILLA

S. M. WALTERS

Botany School, Cambridge University

Introduction

The genus *Alchemilla* (Lady's Mantle) is of special interest to botanists and gardeners, because of the evidence that the different micro-species, which can be distinguished within the two Linnaean aggregate species, have interestingly different histories of spread through human agencies, including agriculture and horticulture. In the very limited time at my disposal it seems more sensible to tell you something of this rather than to attempt a description, or even a catalogue, of the Alchemillas of Britain or Europe.

Conveniently, practically all European Alchemillas (of which some 300 have been distinguished as species) can be grouped into two aggregates: *A. vulgaris,** the Common Lady's Mantle, and *A. alpina,* the Alpine Lady's Mantle. Both these groups are apomictic, reproducing without sexual fusion in the formation of the seed. Distinguishing the different microspecies is not too difficult in most parts of Northern Europe (including Britain), where there is only a limited selection (Walters, 1972). In fact in Britain, 95% of all *Alchemilla* material outside gardens is referable to four species only—the three common '*A. vulgaris*' species (*A. glabra, A. xanthochlora* and *A. filicaulis* subsp. *vestita*) and *A. alpina* L. *sens. str.* If you are interested, therefore, in this critical group, these are the four species to recognize first. It is, however, important to remember that not one of these four widespread British Alchemillas is at all commonly grown in gardens, even in Botanic Gardens.

Alchemilla mollis

The '*vulgaris*' *Alchemilla* which is by far the most commonly grown in our gardens is in fact a much more handsome species, *A. mollis,* which occurs as a native plant in the Balkan and S. Carpathian mountains and extends to Asia Minor and Caucasus. This has relatively large, yellowish (rather than greenish) flowers in which the epicalyx segments are at least as long as the calyx: the large, yellowish flowers and the shallowly-lobed, densely hairy leaves make *A. mollis* an easy plant to identify. It occurs here and there

---

* Authorities for binomials have been omitted for most species; they can be found in the account of *Alchemilla* in *Flora Europaea* 2: 48–64 (1968). For many of the species mentioned they will also be found in Appendix I, in the account of the *Alchemilla* exhibit staged at the time of the Conference.

as a garden-escape, but rarely seems to be effectively naturalized far from gardens. In a garden it spreads vigorously vegetatively, and also fairly freely by seed.

Alchemilla glaucescens and A. sericata

The next-commonest garden Alchemillas belong to the much neater and dwarfer group called the *Pubescentes;* these make attractive rock-garden plants with their silvery-hairy leaves and flowering stems. Among them is our native *A. glaucescens,* which is locally common in the open, sheep-grazed limestone pastures of the Ingleborough region of NW. Yorkshire. This grows very well on light calcareous soils in gardens anywhere in Britain, and in my experience establishes itself more freely from seed than any other *Alchemilla.* In spite of this it rarely persists as a garden escape outside the very restricted areas in N. England, NW. Scotland and Ireland where it occurs wild. In this group is the Caucasian *A. sericata* (incl. *A. rigida*), in which the hair-covering is very silky and sub-appressed; this plant is widespread in European gardens and is sometimes recorded as an escape.

Alchemilla conjuncta

A word about '*A. alpina*'. The native plant, locally common in the Scottish Mountains and the Lake District, does not grow well in lowland gardens. Most so-called '*A. alpina*' in cultivation is in fact *A. conjuncta,* which occurs wild in Glen Clova and is certainly native in the Jura and SW. Alps. The ease with which this plant spreads in cultivation makes it very probable that Buser, the great Swiss *Alchemilla* specialist, was right when he said that Don had planted it in Glen Clova. It is certainly recorded as a garden escape from several British vice-counties.

Alchemilla tytthantha

It is easy to see why *A. mollis, A. glaucescens* (or other *Pubescentes*) and *A. conjuncta* are so common in gardens; they combine qualities of horticultural attractiveness with a vigorous vegetative and seed reproduction. There is, however, one other *Alchemilla* which has settled down as a member of our wild flora which seems to have come to us as a Botanic Garden 'weed' rather than as a positively-encouraged garden plant. This is the Crimean endemic species, *A. tytthantha,* first found in Britain (as an unknown *Alchemilla*) by Dr Margaret Bradshaw in Selkirk in 1956 (Bradshaw & Walters, 1961). Buser knew this plant (as *A. multiflora*) in the Vienna Botanic Garden, where it was in cultivation as early as 1887, and Rothmaler knew it in the Berlin Botanic Garden in the 1930's. It seems to have had a quietly successful spread from one European Botanic Garden to another, perhaps travelling as seed in soil attached to other plants; in this way it could have arrived in S. Scotland via the Royal Botanic Garden in Edinburgh. I saw it recently in the late nineteenth century Botanic Garden of Průhonice, near Prague, and also in the much older Botanic Garden of Charles University in Prague itself.

Weed Alchemillas

This game of identifying the travelling 'weed' Alchemillas of parks and gardens is one that can be played on a world scale. There are *vulgaris* Alchemillas as weeds in more or less artificial grassland habitats in Japan, N. America and Australia. In some cases we know what they are (the Australian one is *A. xanthochlora*); but in other cases we have not yet worked it out. I have for example, a Japanese *Alchemilla* with an interesting combination of characters which is troubling me at present! Even nearer home we are still making tentative identifications, such as the *Alchemilla* in the churchyard at Hampden-in-Arden, Warwickshire, and in an old plantation near Nant-y-ffrith, Denbighshire, which we have provisionally identified as the Caucasian *A. venosa* Juz. It seems unlikely that we shall exhaust these *Alchemilla*-problems, expecially as one suspects that the rate of spread of some of the 'weed' species is relatively fast, and they may be cropping up in new places faster than we can identify them!

REFERENCES

BRADSHAW, M. E. & WALTERS, S. M. (1961). A Russian *Alchemilla* in South Scotland. *Watsonia* 4: 281–2.
WALTERS, S. M. (1972). Endemism in the Genus *Alchemilla* in Europe. In D.H. Valentine (ed.), *Taxonomy, Phytogeography and Evolution*: 301–5. London & New York.

ACAENA

P. F. YEO

University Botanic Garden, Cambridge

Introduction

The genus *Acaena* (of the family *Rosaceae*) has been the subject of various errors and misunderstandings which have affected the naming both of the cultivated forms and those naturalized in the British Isles.

In this paper I am confining myself to those groups in which the flowers are gathered into capitula which are spherical or nearly so, and which have 0–4 spines at the top of the receptacle and either one or two achenes. In other groups in the genus the inflorescence is usually elongated and interrupted, and there are numerous spines distributed over the upper half or the whole of the receptacle.

To the European botanist the spherical-headed species of *Acaena* present an immediate resemblance to the familiar *Sanguisorba* and *Poterium*, the Great and Salad Burnet, to which they are undoubtedly closely related. I am here providing an account of that arbitrary selection of species which happen to be naturalized or cultivated in the British Isles. In Appendix III is presented a formal taxonomic revision of these species, together with a key for their identification. Botanical authorities for the binomials used below will be found in this Appendix.

Acaena affinis

This native of the Magellan Region, the Falkland Islands, and other islands scattered right round the Antarctic continent has not been recorded as naturalized in Britain, but it is in cultivation. It has been received at the Cambridge Botanic Garden from two sources. The first was the late Mrs Margery Fish, of South Petherton, Somerset, from whom the plant was received in 1961 under the misapplied name *A. adscendens*. It is a magnificent garden plant, covering the ground with a deep mat of blue-grey foliage, and raising its red-tinted heads early in May. It looks very like the form illustrated in the splendid plate of Hooker's *Flora Antarctica* (t. 96, 1845). The second source was Mr David Walton of the British Antarctic Survey, who gave me three plants raised in 1969 from seed collected in 1968 in King Edward Cove, South Georgia. These look extremely like the first plant but have proved much less vigorous in our conditions, and so far only a single capitulum has been produced.

Acaena macrostemon

This species is very similar to *A. affinis* but the foliage is paler and green-glaucous rather than blue-glaucous. The leaflets are distinctive too with few, large, deep and irregular teeth, usually more or less folded longitudinally. It is native of Chile and Argentina and not naturalized in Britain. The cultivated plant I have studied is a single female clone received at the Cambridge Botanic Garden in 1898 from Canon Ellacombe of Bitton, Gloucestershire. It is not such a good garden plant as *A. affinis* and sends up suckers from underground rhizomes. The capitula and cupules in the plant cultivated appear to reach normal dimensions, but the cupules are empty.

Acaena magellanica subsp. **magellanica**

This species was first described as *Ancistrum magellanicum* by Lamarck in 1791. Bitter (1910–1911) interpreted the species correctly but clouded the issue by describing *Acaena glaucophylla,* which really belongs here. He described it from cultivation and consequently this subspecies is still found in botanic gardens under the name *A. glaucophylla*. It is a native of the coastal regions of Patagonia.

As a naturalized plant *A. magellanica* subsp. *magellanica* is a wool alien, and has been called "*A. adscendens*" or "*A. laevigata*", names which properly belong to the next taxon. In the garden it makes a continuous mat of glaucous foliage close to the ground, and flowers profusely and conspicuously in early May. The leaves are small, the leaflets produce fewer teeth and the sterile side shoots are usually rosetted.

Acaena magellanica subsp. **laevigata**

This subspecies, distinguished by Lamarck in his original protologue as var. β was named *A. adscendens* by the Danish botanist Vahl in 1804 and *A. laevigata* by William Townsend Aiton in the second edition of *Hortus Kewensis* in 1810. In following Bitter's treatment of it as a subspecies, one has the advantage that the name *A. adscendens* does not require to be brought into use, as it would be at specific rank, even though it has been often and incorrectly applied to the plant here called *A. affinis*.

Acaena magellanica subsp. *laevigata* is native to the Falkland Islands and perhaps the Magellan Region. In Britain it is not known to be naturalized. It has been in cultivation since 1790, when it was grown in the Jardin du Roi in Paris. Among the cultivated species it is distinguished by its deep green leaflets (upper surface deeper green and less glossy than in *A. novae-zelandiae*); it spreads slowly and keeps close to the ground, like subsp. *magellanica,* which it also resembles in its massive, short-spined fruiting capitula. The distal leaflets are noticeably cuneate at the base in luxuriant growth.

Acaena 'Blue Haze'

This plant is still a mystery, for it fits none of the numerous taxa which have been described, nor the known hybrids. The plant in cultivation at Cambridge was received from the Royal Botanic Gardens, Kew, in 1956 as "*A. adscendens*", but while it is not that species its affinities do seem to lie with the group. It has thick stems, but with long internodes, and its leaflets recall *A. magellanica* subsp. *magellanica*.

Although its origin is unknown, and no botanical name appears correct, it still needs a name for reference purposes, so Hilliers of Winchester recently christened it 'Blue Haze'. This name is most descriptive of the characteristic blue-green coloration of the leaflets which are bordered with red, set off by its reddish stems and petioles, as well as its tall red scapes and reddish heads. It is one of the best Acaenas for gardens, and flowers in May.

Acaena glabra

Perhaps related to the *A. magellanica* group, this New Zealand species is almost completely hairless, the leaflets are deeply cut, smooth, pale glossy green above and glaucous beneath. It is little known in cultivation and has not become naturalized.

Acaena fissistipula

This plant from New Zealand is a rather nondescript dwarf species. It is reminiscent of a glaucous and less hairy *A. pusilla* and although as effective in covering ground as other species, it has nothing special to commend its cultivation. It is not known to be naturalized.

Acaena ovalifolia

A native of much of the Andean chain from the Magellan Region to Colombia, this species has only comparatively recently become naturalized in Britain. The abundant colony at Fenagh House, County Carlow, is stated to be a garden escape.

In cultivation it forms deep mounds of bright green, scarcely glossy foliage. The leaflets are very large, more or less elliptical, with many teeth, and are silky-hairy beneath. The stems, which are hairy, are comparatively thin and the fruiting heads green with red spines, although with only two to each floret.

Acaena novae-zelandiae

This is the New Zealand counterpart of the South American *A. ovalifolia*. It also forms a large plant with long, bright green leaflets with fairly numerous teeth. However, the leaves are decidedly glossy, though finely wrinkled, and the under sides are glaucous and slightly silky. In cultivation it makes a good, fresh-green ground-cover and in England flowers in late May.

It is the most commonly naturalized species in Britain and, presumably because Bitter treated it as a subspecies of the widely circumscribed "*A. sanguisorbae*", it is called *A. anserinifolia* in Clapham, Tutin & Warburg, *Flora of the British Isles,* and by Valentine in *Flora Europaea,* this being the correct name for Bitter's "*A. sanguisorbae*". Bitter had several subspecies and our plant coincides with his subsp. *novae-zelandiae* which in Allan, *Flora of New Zealand,* is restored to specific rank. This subspecies also occurs in Australia and has probably been introduced from there as well as from New Zealand, with wool shoddy.

Acaena anserinifolia

The true *A. anserinifolia,* also a native of New Zealand, is hardly ever naturalized in Britain. As a garden plant its value depends on the stock grown. Typically the whole plant is tinted with gravy-brown (although this makes it look rather unhealthy). It is also dwarfer than *A. novae-zelandiae* and the upper surface of the leaf (where not brown-tinted) is quite dull and light or yellowish green. Also the leaflets are usually more hairy, with sharper, penicillate teeth and the capitula are smaller. It flowers earlier, in early May. A preferable form also in cultivation has a deeper-cut, more silky foliage and a better developed brown pigment.

Hybrids between *A. anserinifolia* and *A. inermis* are in cultivation. Superficially they resemble a form of the former species but the leaflets are broader and nearly all the flowers have two stigmas instead of one. In fact it may be said to be the only really decorative *Acaena* in flower, for the numerous capitula have conspicuous white stamens and styles. This plant was sent to the Cambridge Botanic Garden by Mr David McClintock from the garden of Mr Michael Noble, Strone, Cairndow, Argyll. Seedlings of this plant show startling genetic segregation with remarkable variation from plant to plant. Some resemble *A. anserinifolia* with brown, red or gold tinting, others are purplish and yet others with pure blue-green glaucous foliage. The seedlings make a delightful, variegated ground-cover.

Acaena pusilla

This New Zealand plant would, perhaps, be better treated as a subspecies of *A. anserinifolia,* which it resembles in minature. The flower heads are only 4 or 5 mm in diameter, excluding the spines. It is only rarely naturalized, being well known from Golspie and one other place in Scotland, and from one locality in Ireland.

Acaena caesiiglauca

This New Zealand species is also a member of the *A. anserinifolia* group. It has not been recorded as naturalized in Britain but forms a most attractive garden plant, vigorous, but not overwhelmingly so, and with beautiful blue-hirsute foliage becoming lightly purple in the autumn.

Acaena microphylla

This native of New Zealand has not yet become naturalized in Britain but in cultivation, where it is frequently met with, it forms a mat close to the ground with brownish colouring like *A. pusilla* and *A. anserinifolia,* but the leaves are minute with numerous leaflets, evenly graded in size. It is mainly grown for its abundant and ornamental fruiting heads which are covered in thick, scarlet spines, and may virtually conceal the foliage; the spines, which in this species are soft in texture, are without barbs.

Acaena inermis

In contrast, this species, also from New Zealand, has no spines on the fruit (whence the specific epithet), but its dwarf habit is similar. The stems are stolon-like, and, rooting immediately, spread rapidly. The leaves are a dull greyish or purplish colour, which virtually camouflages it against most kinds of rock and soil.

Presumably because of its spreading habit, it has several times become naturalized.

Acaena buchananii

Acaena buchananii is likewise a dwarf-growing New Zealand species but is distinctive on account of its unique lichen-green foliage and few-flowered capitula hidden beneath the leaves. It makes a very neat, dwarf ground-cover.

Opportunities for study

For the botanist the species of *Acaena* provide opportunities for observing the introduction and spread of newly arrived plants. For the gardener the genus may have limited usefulness but the species are valuable for the restricted purpose of ground-cover and as an outstanding example of natural variation in colour and texture of foliage.

It is my hope that in this genus the "twain may meet", that is the botanist and gardener, at least to the extent of adopting a common nomenclature. I think too, that one can also expect to find the two parties viewing *Acaena* from closer standpoints than they do in those genera where the flowers are showier. As an afterthought, I also hope that two other parties may get together, in this case the botanists of Australia and New Zealand, to compare and correlate the species common to their two countries.

THE ROLE OF NURSERIES

WILL INGWERSEN

What the nurseryman can offer

Many of our British wild flowers have a place in gardens, and this has led to the partial extinction of a number of species, quite apart from the inroads made upon native stands by road construction and the development of new towns and other activities which encroach upon our diminishing countryside. It is of the utmost importance that the collecting of British plants from their native habitats should cease.

The role of the nurseryman in anything which has to do with the preservation of our shrinking population of wild flowers is an important one. The more so since the burden of what should be done will inevitably fall upon fewer and fewer shoulders. This is due to the regrettable, but I suppose inevitable, tendency for nurseries to grow 'more and more of less and less'.

With the increasing number of only semi-horticultural retail outlets the situation is approaching when a few producers, working on a very large scale and using mass-production methods, will produce the great majority of plants which gardeners can buy. Production on factory lines does not leave room for the cultivation of off-beat plants in necessarily small quantities.

It will, therefore, fall to the lot of a few specialist growers, such as myself, to do what is possible by way of propagating those British wild flowers which are of garden value. This will not only ensure their survival, but will enable gardeners who like to stray away from the standard varieties of garden plants, to buy less usual kinds and to play their own part in conservation and preservation. Mr Thomas is to speak to us later on the role of the private gardener in this context.

Of course, even a dedicated die-hard such as myself, has to retain some commercial sense, or I too would cease to survive. Obviously I could not afford to produce quantities of plants for which there would be no demand. Botanical rarities and plants with no appeal to the gardener would be the concern of Botanic, and possibly Municipal Gardens, and this is an aspect we shall be hearing about from Mr Henderson.

British plants nurseryman can grow

At this point I would like to offer a short list of names of some of the plants in which I believe nurserymen, able and willing to do so, can take a legitimate commercial interest. Plants in this context which come immediately to mind are—

Aconitum anglicum	*Dianthus gratianopolitanus*
Actaea spicata	*Dryas octopetala*
Ajuga genevensis and	*Gentiana verna*
A. *pyramidalis*	*Geranium* spp.
Alchemilla conjuncta	*Helleborus* spp.
Anagallis tenella	*Linnaea borealis*
Anemone pulsatilla	*Minuartia verna*
Arbutus unedo	*Polygala calcarea*
Arcostaphylos uva-ursi	*Potentilla fruticosa*
Betula nana	*Primula farinosa*
Butomus umbellatus	*Saxifraga oppositifolia* and
Cornus (Chamaeperi-	S. *aizoides*
clymenum) suecicum	*Silene acaulis*

Excluded from this list of a score or more of names are the important groups of ferns, orchids, grasses, bulbous plants and heathers, which are too complex to be dealt with in this short session, but which all contain many kinds worthy, not only of survival, but of cultivation.

The duty of any conservationist, as I see it, is to cultivate the best selected clones and forms of our wild flowers. Of some of them of course, such as *Dryas octopetala, Gentiana verna* and *Saxifraga oppositifolia,* the plants in cultivation, more often than not, originated from stock which came from stations outside the British Isles.

Not only will nursery-grown plants of our wildlings be made available to those who, wishing to grow them, very properly hesitate to collect them from their native habitats, but they will also form a reserve in the event of a local and total destruction of a species which it might be desirable to re-establish.

DISCUSSION

Dr A. J. RICHARDS asked what role nurseries could play in providing stocks of native trees for introduction in areas which needed reclamation, particularly in tree planting along motorways. It seemed to him that most of the stocks available consisted either of cultivars or of foreign species. These areas—which seemed now to be potential new nature reserves—were not being planted up with native species nor with native stocks of trees. He wondered if it would be possible for Mr. Ingwersen to comment firstly, upon which stocks existed in nurseries and secondly, whether it would be possible for them to be provided in quantities suitable for planting by the Ministry of Transport, local authorities and so on.

Mr INGWERSEN replied that he was not a tree nurseman and wondered whether he could pass the question to Mr. Hillier, who was much better qualified to answer it than himself.

Mr H. G. HILLIER replied by pointing out that we had extremely few native trees in the British Isles, but felt that the major part of the problem had to do with nurserymen knowing in advance what stocks were required. He was sure that if nurserymen knew what was to be wanted they could produce all that was required. However, a sudden order for, say 100,000 *Acer campestre,* or something like that, at a certain size, would be a bit difficult for anyone to meet. They could be grown and then wasted for lack of demand. He thought it important that the Ministry concerned should let nurseymen know what they have in mind to plant during the next five or seven years, for then the nursery trade of this country could produce all that was wanted. But with such few native trees there would not be a very big selection.

Mr R. L. GULLIVER asked whether there was any possibility of getting shrub species other than hawthorn, perhaps blackthorn or some other native species, for planting in hedgerows beside our motorways.

Mr INGWERSEN replied that this question really had the same answer as the previous one. Our nurseries, were completely capable of producing these plants in quantity, but they could not do so at short notice.

Mr D. M. HENDERSON commented that whilst the subject concerned nurseries he would like to make a contribution. When he had been up the hills in Scotland looking at the very tenuous populations of *Salix lanata,* in the past he had always thought that nurserymen must have been there and pinched cuttings or taken roots, and what a shame that was. However, in the last six or seven years he had realized that he had been completely wrong. He now believed we should bless the good nurserymen who had taken a few cuttings. Nobody in their senses trudged the whole way up to take cuttings today. He thought of *Salix lanata* now lined out in certain nurseries where one could buy it very easily, and *Betula nana* was probably another very good example. He considered that some nurserymen had contributed tremendously to the conservation of a few of our plants. In particular, he felt that it was plants which could be propagated vegetatively which had been conserved by this means by nurserymen, and he thought this was a movement which one should do everything to encourage.

Mr INGWERSEN believed this to be very true. He thought the nurseryman's role important because material for propagation should be taken by people who were qualified to deal with it. It was not much good a tourist taking some cuttings of a plant, throwing them into the back of their car to take home, and then a week later putting them in the ground and hoping they were going to root. Cuttings had to be handled skilfully and carefully, and by careful pruning and the removal of seeds and cuttings, while leaving the original stand in good condition, the nurseryman could play his part in the conservation of desirable native plants.

THE ROLE OF THE PRIVATE GARDEN

GRAHAM THOMAS

The National Trust

Gardening through the centuries

During our early history and through mediaeval times botanists and gardeners were one and the same thing. Plants were grown mostly for food or for their curative powers, factual or supposed, and so far as we know in no particular order or design.

The two pursuits became separated more and more as time went on because botany gradually became a true science and gardening developed into a craft. Gardening became further divided in itself as it developed two allied though diverging ideals: the culture of plants for economic purposes or for their ornamental effect.

During Elizabethan times these two styles had scarcely emerged, though we do, I believe, first hear of the professional architect then and architects have always had a hand in garden design. Though gardens were designed with arbours, parterres and the like, I think I am right in saying that apart from the obvious uses of climbers on walls and arches, the plants were grown in general mixture, with the stress on plants rather than their effect together.

The Stuart and Georgian periods were quite different. The former became formal and very inward-looking with hedges, alleys and parterres composed of evergreen and coloured gravels, and the latter informal, seeking to create idyllic landscapes with the aid simply of grass, trees, water and often buildings of classical persuasion. This is at least what happened in the large gardens of the rich, the fashion-setters; the plants and flowers were left to the kitchen gardens and of course to cottage gardens. (It is indeed the kitchen and cottage gardens which have preserved through the centuries the numerous treasured plants which today we are still re-discovering). During these two periods—roughly for 200 years—it will be realized flowers played little part in the great gardens.

It is easy to label these long periods of horticultural development in this way, though the whole matter is much more involved than I have made it appear. In Victorian times, flowers and plants came into their own again, from then and up to the present day a vast complex of garden styles has evolved. This has been in great part due to the tremendous influx of plants from all regions of the world, both hardy and tender in our climate. Another major factor was the discovery of the use of glass structures for rearing tropical plants, coupled with the cheapness of coal and garden labour. Thus arrived the popularity of 'bedding out'; also the beginnings of arboreta to contain the conifers, Rhododendrons and other trees and shrubs from all parts of the Temperate World. Later on, ferns and herbaceous plants received their due, with alpines reaching popularity in the early part of this century. Today there are ever greater extremes practised in the pursuit of collections of plants and also in garden artistry.

Under glass, plants had a more chequered career than outside, owing to economic upheavals, rising costs etc. Vegetables and bedding plants were continually 'improved' by hybridists, and fruit went more slowly through the same process. Therefore, from our point of view today, I think it best to concern myself with looking into the part played by hardy ornamentals, bearing in mind throughout that it has been the botanically minded gardener who has always created the greatest collections of plants. The artist-gardener achieves his effects with fewer materials.

The role of private gardeners

Let us look into the part played by private individuals during these 400 or more years of gardening—creating their collections by purchasing, begging, exchanging and even stealing (I mention the last means because of the pilfering from which we suffer so much in the National Trust and also because it is amusing to contemplate those pilferers who put their stolen cuttings and seeds under their hats or into their umbrellas: meeting a lady or a sudden shower can have embarrassing results). In addition individuals have nobly supported expeditions abroad or have braved the wilds themselves in the endeavour to send plants home to enrich our gardens.

No wonder that these islands contain thousands of smaller gardens and hundreds of bigger ones, providing overall a vast reservoir of plants, both species and varieties, for the enjoyment of and study by gardeners and botanists. I was looking through my bookshelves the other day when preparing this text and came across several bound volumes of great private collections of plants. That at Rostrevor, Northern Ireland, in 1911 listed about 2600 trees, shrubs and plants; in 1932 the Borde Hill book enumerated approximately the same number of trees and shrubs, while at Wakehurst, another Sussex garden, over 5,000 were listed in 1942. But let us not forget the smaller gardens, yours and mine, where perhaps 1–2,000 species and cultivars may be grown. It is astonishing what can be grown in just one acre—without spoiling the effect and making merely a jumble of plant forms.

We are fortunate today to be able to visit two exceptional collections —the thousand upon thousand of trees, shrubs and plants grown in Norfolk by Mr Maurice Mason, and Mr Hillier's arboretum at Jermyns, Winchester, an area of 85 acres housing another astonishing collection. These two, together with the Savill Gardens, are possibly the most extensive collections to have been formed in recent times. You may well ask how it is possible for individuals in these days of constantly rising costs to maintain such huge areas satisfactorily. The answer is by making the design simple, and by tailoring the garden to the use of machinery. In fact I think it is probably cheaper to maintain the large simple garden today than it has been for 60 or more years. Collections come and go with the lives of the enthusiasts who direct them. An owner-gardener may spend 10, 20, 30 or more years accumulating his collection of plants, but as likely as not it disintegrates after his death. But meanwhile the true gardener has given away plants, cuttings and seeds and has enthused countless others, who in turn build up their collections, each with different tastes and outlook, in different soils and climates. The general trend is towards more and more specialization in smaller gardens. You would not need a big garden to grow, for instance, an exhaustive collection of Violets, species and cultivars, or old Auricula cultivars, or the new bulbs from the Middle East.

This constant process of building up collections of plants is very much subject to fashion and economics, which are both closely related anyway. We can recall the tulip craze of the 17th Century, and the fern craze of Victorian times, both short-lived; or the rose craze which is still very much with us, having been started by the Empress Josephine, Napoleon Bonaparte's first wife. In the early part of this century a sudden great interest was taken in the breeding of Montbretias, cultivars and hybrids of *Crocosmia,* and it is one of my pleasures today to try to bring together the hardy named varieties. I have managed to pick up some half-a-dozen from gardens in this country. Many of you will remember the exhibits in this hall in the early 30's when stands had 50 or more species and varieties of Kabschia Saxifrages, and I suspect it will not be long before some enthusiast with an eye to plants of sterling, all-round garden worth will begin to collect together once again the best compact Irises of the first 30 years of this century. Plants and genera may lose favour, sometimes owing to boredom, or conversely their popularity may be delayed for many years, as with the Hybrid Musk roses which were bred before their time.

Every garden, whatever its size, has something of special interest —something growing better than you have ever seen before, the largest specimen, the highest quality, the old, the rare or the new. Queer things crop up in unexpected places. Only this year in two widely separated hotel gardens, in the north and in the west, I came across plants seldom seen. In one was *Boltonia latisquamata* and *Lonicera xylosteum,* in the other was *Crocosmia pottsii.* None of these can be called anything but rare, and yet the hotel gardens had no standing horticulturally, whatever there cuisine may have been. The moral of this is never to conclude their is nothing worth seeing however poor the garden appears to be, and the lower the state of horticulture the less will a cutting be missed!—though I must mention I did not 'pinch' any of these plants.

The gardens of the National Trust

Let us now see how the gardens of the National Trust fit into all this. The Trust owns far more gardens than have ever been under one management before. And you would not I am sure expect me to say much about the Trust without making my usual remark that it is a private charity, not a Government department; a *needy* private charity into the bargain. Whatever its shortcomings or disadvantages it has one great advantage over private individuals, and this is continuity of purpose. This applies very much to gardens, as you all know. I have the honour to be its Gardens Adviser and like to think that there will always be a Gardens Adviser to act as an interpreter of its main policies, and to guide its head gardeners along an inherited line of thought. It owns gardens enormously different in styles. Apart from certain arboreta, its gardens act almost always as an embellishment to a house. Plants are enjoyed for their own beauty to a greater or lesser degree in each garden, with least stress on them in the great formal and informal landscapes. In these, of course, a collection of any kind of native or exotic plant would be out of place. In some gardens, on the other hand, the Trust has inherited from an individual, or generations of individuals, collections of plants enshrining the tastes of the periods concerned. Just as these owner-gardeners would become interested in one or more genera, so the Trust feels that, in appropriate gardens, it can well 'become keen' as it were on a genus, or set of plants, in its own right. This all helps to keep our gardens separately and differently interesting.

Let me tell you of a few examples. No doubt a trained botanist could make a most interesting list of native plants growing in Trust properties; I will just hint at a few in our gardens. On the lawn at Arlington Court, Devon, for instance are to be found *Wahlenbergia hederacea* and *Anagallis tenella,* while in the woodland rides *Claytonia sibirica* and *Primula japonica* are naturalized. Very soon now, after rain, the Ladies' Tresses orchids will be spearing through the lawns at Berrington Court, Herefordshire, and at Uppark, Sussex. At Wallington Hall, Northumberland, recently I came across *Pyrola rotundifolia* in the garden woodland, and both *Asarum europaeum* and *Cardamine trifolia* are naturalized nearby. Species of *Epipactis, E. latifolia* and *E. purpurata,* are to be seen at Lyme Park, Cheshire, and Benthall Hall, Shropshire, respectively. It will be obvious from the natural growth of these plants that we use weedkillers circumspectly.

We have inherited a number of special collections, among them being some crocuses at Benthall Hall, originally planted by George Maw who lived there and wrote his *Monograph of the Genus Crocus* in 1886. At Killerton, Devon, there is one of the earliest planted arboreta in these islands, containing many conifers and Rhododendrons; Rhododendrons also figure largely at Trengwainton, Cornwall, where many species derive from one valley in China whence seed was sent home by Kingdon Ward in 1929. There is a notable collection of hardy ferns at Sizergh Castle, Westmorland, planted in 1926, and some named cultivars of walnuts at St. John's Jerusalem, Kent, to which we intend adding. And of course there is the great collection of autumn-colouring trees and shrubs at Sheffield Park, Sussex.

Individual plant rarities in our gardens are too numerous to mention more than briefly. The pale blue form of *Gentiana asclepiadea* seeds itself mildly at Sizergh Castle, *Narcissus pallidus praecox* (Parkinson's "Early Strawe-coloured Bastard Daffodil") is naturalized at Killerton, and one of the three original clones of *Aster frikartii*, named 'Mönch' (the other two were 'Eiger' and Jungfrau'), grows well at Blickling, Norfolk. This is a rare plant, and far better than 'Wonder of Stäfa' which seems to have supplanted it in gardens.

As I mentioned earlier we seek to extend these collections and to add to them, and also to plant differently in each garden. For some years now we have been assembling all species and cultivars of Hydrangeas at Trelissick and are concentrating on Magnolias, Hypericums and Deutzias at Lanhydrock; both gardens are in Cornwall. Several species of Ash have been started at Arlington Court, Devon, and we have set aside Alders for Farnborough Hall, Warwickshire. Some of you will know of the large collection of old and species roses which I built up a few years ago at Sunningdale Nurseries; the entire collection is being duplicated in the beautiful walled garden at Mottisfont Abbey, Hampshire (it has already been planted also at the Royal National Rose Society's display garden at St. Albans, Herts.) These are just a few examples of our work in the conservation of garden plants.

Conclusion

At times I have thought how wonderful it would be if the work of botanic gardens in making collections of species of different genera could be extended throughout the country. We cannot expect every genus to grow well at Kew for instance, nor could any one garden contain all the species and most of the cultivars of all genera. I visualize gardens of permanent duration—the Public Parks, certain gardens of the National Trust, the Horticultural Institutions, Experimental Stations and the like, each being responsible for maintaining all known species and most cultivars of one or more genera. I wrote a letter about such a scheme to the *Gardeners' Chronicle* some years ago, but it elicited practically no response. The difficulties are many. There would have to be some central control, otherwise, with the coming and going of superintendents and curators, the collections might well deteriorate from time to time. Maps would have to record the plantings, to avoid labels becoming misplaced. Many other points occur, but I very much hope that one of the results of this conference may be that we shall work towards something of this sort. There are, however, two sides to every question and some of you may well say that one of the great joys and interests of assembling plants in your gardens is the excitement experienced when one comes across a rarity unexpectedly. But there would still be plenty of scope for this sort of collecting, even if the main genera were each given special homes. Fortunately, whatever may happen, it is my experience that few absolutely first class plants become lost to cultivation; by first class I mean good in every way, including health and longevity.

I suggest that the greatest contribution that private gardeners like ourselves can make to the world of botany and gardening is to go on assembling and growing all kinds of plants—if you have a mind to do so, and belong to the collecting sort of garden enthusiasts rather than the artistic ones; by so doing you will help to keep the best sorts in cultivation and by sharing your plants with others you will help to keep rarities in circulation. This is the main point; always propagate and distribute a rarity if possible. Who knows— your plant may die and you will be glad to have a piece of it back some time.

DISCUSSION

Mr R. L. GULLIVER asked whether, in view of the fact that a lot of land was now in the hands of local authorities and the Health Service, and in the face of vandalism, work study schemes and a shortage of labour, it would be a good idea for people to pressurize their local authorities to have the sort of planting done that Mr Thomas advocated, rather than settling for the nondescript shrubs and grass of the type one seemed to be seeing more and more.

Mr THOMAS agreed that it would be highly desirable for some public bodies to show more horticultural interest, but wished to comment that the National Trust is not an advisory body. It could open its doors so that people could see what was going on in its gardens and the people that came might very often disapprove of what was being done, for the Trust did not hold itself up as doing everything correctly, yet he believed it was only through example and study that standards could be improved. However, it should be said that in the public parks and areas under municipal control very much larger numbers of people, by and large, had to be coped with than in the properties under the Trust, nor did their land have so much resting space in the winter.

In addition, Mr Thomas remarked that there was one question he had wanted to raise earlier, in connexion with the conservation of British species. Had this gathering anything to offer by way of advice to County Councils not to spray minor country verges? Earlier in the year he had been in the North Riding and there, in May, they were already pouring weed-killer on to the poor little patches of Cow-parsley along small lanes in the countryside.

Miss C. M. ROB commented that from her knowledge the spraying in this case was not done by the local authority. The North Riding were extremely good, and in this matter, the County Surveyor did very much as he was asked. It was the local farmers who tipped out the residue as they returned from the field rather than take it back into the farmyard in their sprayer.

DR F. H. PERRING remarked that he thought it ought to be said that Dr Way of Monks Wood Experimental Station had that summer visited every County Surveyor in the land, as far as he was aware, and had spoken to them all personally about this problem of spraying roadside verges. He believed that it was very unlikely that more than a very few would resist the tendency towards cutting rather than spraying roadside verges. He felt that in most cases where spraying may be found to have occurred next summer it would nearly always prove to have been private individuals at fault and not the County authority.

Mr D. McClintock commented that on the subject of collections of plants, there had been built up over the last three or four years at Harlow Car, in Harrogate, as complete a collection as could be made of all the Heather cultivars. The object had been partly to collect as many as possible to compare them and also to try to find out which were mis-named and try to get their correct names. The aim was to collect every Heather cultivar, and anybody who could contribute further to the collection was most welcome. At the moment they were growing about 450 or 500 Heathers and there was a desire to make sure that the collection was complete.

Mr R. D. Nutt said that he was delighted to hear of Harlow Car and the collection of Heathers, but having collected one or two genera himself, his own view was that it would inevitably be the private gardener who would conserve and hunt around for different forms, not the Botanic Gardens. The real problem, he was afraid was the muddling up of labels, as well as the care and attention that such collections needed, and they did require an enormous amount of attention in order to maintain them properly. So, while he would support, and in fact encourage, local authorities and botanical institutions to make collections, he did not feel that this could be the ultimate answer.

Mr McClintock added that the Heather collection was not in fact supervised entirely by the staff at Harlow Car, but by a sub-committee of six from the Heather Society, which watched the points Mr Nutt had mentioned kept the records, acted as a permanent body to look after the collection and gave some continuity.

THE ROLE OF BOTANIC GARDENS IN HORTICULTURE AND FIELD BOTANY

D. M. Henderson

Royal Botanic Garden, Edinburgh

A variety of roles

Whatever the role of botanic gardens may be in the future, their main and continuing function has been the storage of large, living, plant collections. And the collections have been used for some purpose—general education, research, medicine, reference. To a great extent the nature of this usage has been determined by an associated institute—a university botany department, a medical school, a research herbarium. The university gardens supply teaching material and specialized research collections, the gardens with taxonomic institutes supply living plants for research and allow the taxonomists to become familiar with the whole plant, not just what can be preserved on the herbarium sheet or in the pickle jar. Implicit also in the idea of a botanic garden are that the plants should be labelled and that the collections should be arranged in some order—for aesthetic appeal, for biological interest or along taxonomic lines.

Where does the topic of this Conference—Horticulture and Field Botany—fit into this picture? It would be simple to extract a string of single examples showing how individual gardens have made isolated contributions to horticulture and even to field botany. I would agree that although there is indeed some connexion with horticulture, over the wide field, the relationship with field botany is tenuous. In parenthesis I should add that in the context of a joint conference with the Botanical Society of the British Isles I take field botany to mean field studies mainly of the British flora and not botanical surveys overseas. It is idle to pretend that botanic gardens can be concerned with a great range of distinct disciplines. Botanic gardens, just like other organisations, must specialize. In previous centuries they were the only centres, not only of general botany and taxonomy, but of forestry, plant physiology, plant pathology and so on. Botanic gardens often cradled these disciplines which now merit separate institutes or entire government agencies. By their loss gardens have specialized. I believe this is a process still in operation and in a sense botanic gardens have returned to their original aim—the utilization of living plant collections.

The living plant collections

The quality and origin of collections in botanic gardens circumscribes their use. I should remind you how they are amassed.

The plants may come directly from the field—from plant collecting expeditions. Their origin is known and the plant represents the plant in nature with some degree of accuracy. Or they may come secondarily—by gift or exchange with other gardens. At most we estimate that we have some record of the wild origin of 10 per cent of the Edinburgh collection of some 15,000 species; the remaining 90 per cent are from secondary origins. This is the nature of the plant resources available to horticulture.

There is an unceasing movement of plants through this system, so botanic gardens have always played a major part in trying to introduce new plants for eventual release to general horticulture. The system has been haphazard in many respects. The scale of the operation has always precluded careful comparative tests, although the cynic has said that survival in a botanic garden is the best test of a plant's toughness for general gardening—and there is more than a grain of truth in this. The botanic gardens' role in plant intro-duction has been to grow and pass on foundation stocks to commerce, research workers and serious amateurs. The whole system of holding and supplying less common plants may come under stress in the future. Commercial horticulture is now specializing to such a degree that no longer do nurseries stock a large range of plants—the uncommon has become much more difficult to find in catalogues. Clearly the large diverse collections of botanic gardens—and let me add of private gardens also—are going to be more important as sources of the less common plants. One of the chief glories of British gardening is the diversity of plants used. The general trend in commerce is towards large-scale use of a few subjects. Where, one may ask, is the wide range of stock to be held in the future? Botanic gardens surely have a greater role to play here than in the past.

The collections of gardens also provide for research—parents for plant breeding, for horticultural taxonomy and so on. The provenance of the collection is then important. Botanic gardens now realize that it is important to have properly documented collections. To improve documentation the quality of exchange material requires to be improved. The enormous seed lists of ill-named, mis-named infertile hybrid seed of doubtful origin are too casual and wasteful of effort. They also serve horticulture ill. Smaller exchanges of high quality stocks will be of greater value.

Specialization

There is also another sense in which gardens must specialize if they are to be successful in the future and serve the many other disciplines which may seek their help. Individual gardens must specialize to a greater degree in the plant collections they hold. I would suggest that gardens can fulfil their various roles best by holding a carefully chosen general collection, sufficient to give an overall picture of the plant kingdom (certainly not just as large a collection as they can lay hands on), then a limited number of specialized collections. The number of these collections will depend to some extent upon space and almost certainly upon the staff resources available to care for them adequately. We should look forward to the day when the boast of a garden is the quality of its specialized collections, not just the total number of plants grown. Such collections would also ease a problem of direct concern to horticulture which I have just mentioned—the mis-named plants and hybrid seed which have been exchanged in the past. Smaller general collections have advantages as well. Museums now tend to display carefully chosen examples and explain them adequately. Gardens have only begun to explain their treasures to the general public. Surely there is much to be said for the explained, selected collection rather than the diffuse jungle of the Victorian planthouse.

So far I have dealt at some length with the problem of botanic garden collections at the non-commercial level. As I have hinted, they may serve as storehouses for uncommon plants in the future. Should they do more? Are there many plants in botanic gardens whose merits should be promoted? Would it be the legitimate function of a garden to do this? It must be borne in mind that promotion of a good plant is useless unless a stock is available. Perhaps eventually we should think of a 'nuclear stock scheme' for good plants, for their multiplication for distribution on any scale is beyond the capabilities of gardens at present. Cognate to the supply of 'good' plants to horticulture is the search for new strains, the testing of individuals of a species from different areas. If the experiences of the provenance trials of the Forestry Commission have anything to teach horticulture it is that casting the net wider for other provenances or ecotypes might convert our ideas on the behaviour, especially hardiness, of a species in cultivation. Our amenity plants have often come from a single introduction—B. K. Boom noted some years ago that all *Ligustrum ovalifolium,* the common Hedge Privet, stemmed from one original source. Is this testing of provenances a task for botanic gardens? The answer is perhaps, yes, when the plants are part of the specialized collection of the garden, but surely as a general problem it is more the realm of a research station with responsibility for amenity horticulture.

In this survey I have suggested fairly strict limits on the place of botanic gardens in relation to horticulture in general. I would be even stricter as regards field botany. True, botanic gardens cultivate a few plants of wild British origin as decoratives and the British collection can be an interesting part of a demonstration garden, but for serious study of variation in British plants the botanist requires an experimental plot, *not* a botanic garden. At this point specialization impinges again, some gardens should have major specialist collections of British plants, others only a general selection. But I would stress that the maintenance of such a collection requires *permanent* specialist staff. We have had several

British research collections in Edinburgh over the last twenty years but when research workers moved the critical control was lost, and so also was the collection. The lower the taxonomic rank of the items of the collection the more important is specialist supervision; ecotypes are usually even more difficult than species. The relatively permanent staffing of national botanic gardens might suggest that they could better care for critical collections, for a long-term commitment does not accord well with the staff changes and altering research priorities, often characteristic of a university-based garden. But the national gardens are more orientated towards extra-British plants. I often suspect that collections for specialist research can be maintained only under the supervision of the specialist and that when the research stops long-term maintenance is a vain hope.

I hope I have been provocative enough to stimulate discussion, for the roles of botanic gardens do require discussion. They have long been accepted rather uncritically. My main thesis is clear, to be successful botanic gardens must evolve, they must be relevant to the times and to do this they must specialize. They must develop specialism in showing the plant world to the visitor, and, with advantage, specialism in their plant collections. Do not let us pretend that today they can be major general centres for horticulture and field botany.

DISCUSSION

Mr R. L. GULLIVER remarked that one of the growing roles of botanic gardens was public education, and asked how great a number of plants from different habitats this required—and the specialization in this connexion—and what was being done to encourage the use by school parties of the gardens' facilities?

Mr HENDERSON replied that most of the existing gardens in Britain, as they stood at the moment, were so full of plants that he did not think there was any room, without radical changes, to make major ecological plantings in them. He thought this was one answer, as it applied to the existing situation. For new gardens, there seemed no doubt that they could be planted as ecological units and he believed that Mr David Sayers was in fact involved in doing something of this sort at Durham at that moment. However, he stressed that it was a difficult thing to superpose ecological plants on present gardens which were already over-stuffed with plants.

Mr J. S. L. GILMOUR, while agreeing in general with what had been said about specialist collections, wondered what Mr Henderson had in mind as regards the financing of such collections, if they were not under any of the existing botanic gardens, either university or national. Had he an idea that there might be some special form of finance, say from the Government?

Mr HENDERSON replied that he had not, but thought that, if we were to get our "houses" in order, and divided out the responsibilities between the gardens in Britain, we could go a very long way towards maintaining these specialist collections, perhaps without large extra resources at first. He felt that the problems were duplication, and all the other ills that botanic gardens suffered from. In addition, he pointed out that implicit in all this was that we know what one another really had. Many gardens did not yet have a full inventory of their holdings. He was happy to say that in Edinburgh they were just checking the first print-out of the entire collection of the garden. They had had a card index for a long time, but would now have a list of what was growing. This would help in the assessment of what they were going to do with their collections. He hoped that they were thus moving in the right direction.

Dr S. M. WALTERS agreed that the documentation job was the most important thing before one moved towards the goal of specialization. But would not the second stage be to ask the question, "What use was the 90 per cent of material of unknown origin in the botanic garden". Then the next question, "Could we not, in fact, throw away 50–75 per cent of what we had, and establish relatively quickly more useful collections which were more in tune with the demands which science was making on the gardens at the present time?"

Mr HENDERSON believed that this was correct. Once one had an inventory of what was in the gardens, then by discussion and collaboration one decided what to keep. One had only to look round our glasshouses to see tropical forest trees pot-bound in pots about a foot across, yet there was no space, in fact no garden had sufficient space to grow these trees to their proper size. The matter was out of proportion now for, with rapid air travel, there was no need to hold some of these plants in this country at all. The burden of conservation throughout the world could be spread to the botanic gardens *in situ*. As an example, one could take the case of Mauritius and its botanic garden: with only about one-eighth of the island left in native forest one had an interesting endemic flora becoming extinct, yet hardly any of the native plants grew in the Pamplemousses Botanic Garden, founded about 1840. Nevertheless, it would be relatively simple to put the native trees into Pamplemousses. But there were much greater problems in other parts of the world and, for instance, trees were much more easily conserved than herbaceous material.

Dr W. T. STEARN said that he would like to make a plea about botanic gardens not throwing away their "rubbish". Mr Bowles and he had been concerned for many years with the genus *Anemone,* and Bowles never in his lifetime could remember seeing the original *A. japonica.* It had been thrown away by botanic gardens as "rubbish". However, it happened that he had been in the Scilly Isles at Tresco when they were throwing out an *Anemone* which was "rubbish" and it just happened that he realized what it might be—it was in fact the very last stock in all Europe of the original *A. japonica.* This species was of very great importance in the evolution of the so-called Japanese Anemones that we had in gardens. He hoped he would be forgiven for telling the story of the Japanese Ambassador who visited Sir Frank Crisp's Japanese garden and said, "Wonderful, we have nothing like this in Japan"! Until very recently this comment was also true of the "Japanese Anemones" of gardens. They were all evolved by hybridization in the Horticultural Society's Garden at Chiswick.

The importance of botanic gardens not throwing away "junk" was that very often they did not realize that they held the one and only stock in cultivation anywhere. He mentioned three plants: *Colchicum callicymbium,* which he had described as a new species; we still did not know exactly where it came from in the wild state, and it was now extinct in the one botanic garden that had it. *Tofieldia calyculata,* which he had also wanted for some recent purposes; it was now dead in the one botanic garden that grew it. *Milula spicata,* if we wanted it for analysis purposes we had to go to Tibet to get a new collection from its wild localities. There were a number of other instances of this sort.

Further, he wished to emphasize the importance of the private individual. If one had travelled on a London 'bus between 1920 and 1930 one might have met a very amiable conductor named Bates who had a small house in Hounslow. His garden was only from here to the back of the hall, yet he had the largest collection of succulents in private hands. Whenever N. E. Brown described new species at Kew, he handed them over to Bates because he was certain that his was the one place where they would be kept alive. In due course,

Kirstenbosch in South Africa had to send to this London busman to get back some species which they found had become extinct in South Africa. Bates' private collection was purchased by the University of California. He believed that this was a very good example of how important an amateur could be in maintaining a collection intact.

Mr HENDERSON replied that he wanted to make one point clear. One would only throw plants away after careful checking and if there had been greater concentration on doing things well they would probably not have lost the *Milula*.

THE RELEVANCE OF GENETICS

KEITH JONES

Jodrell Laboratory, Royal Botanic Gardens, Kew

Introduction

Modern genetics is a sophisticated and sometimes complex science which informs us about the nature of the control of plant variation and its inheritance. It can for our purposes be divided into two main aspects, namely, a) the strictly genetical analysis which deals with gene effects on character production, type of gene interaction and segregation and, b) the study of chromosomes which carry the genes and which exert a pronounced effect on patterns of inheritance. These two disciplines are, of course, closely interrelated and the results of one are meaningless unless considered with regard to the other—biologically they are inseparable. If then genetics gives us an understanding of the control of heritable variation it follows that it might be highly pertinent to all who attempt to manipulate this variation for their particular purposes. This applies as much to the ecologist, physiologist and taxonomist as it does to the evolutionist or plant breeder, whether in the agricultural or horticultural fields. In particular, I will attempt to demonstrate some of the ways in which horticulture has benefited and will continue to benefit from the applications of genetics, but by way of introduction I will first consider the debt which genetics owes to horticulture. In the Royal Horticultural Society's hall this, I think, would be most fitting, and indeed just, and will enable us to approach our subject in the realization that science progresses as a result of the continuous interaction of many disciplines.

The horticultural background of genetics

I suppose that by a logical process of deduction we could consider the Garden of Eden as the horticultural paradise in which the first genetic experimentation started, but it is another garden where our science really began. It was the diminutive garden of that Benedictine monastery in Brno where Father Gregor Mendel commenced his studies on the edible pea around the year 1857. Using this plant—itself the product of man's influence on nature—he was able to demonstrate that inheritance is a particulate affair. That is to say factors which controlled the characters of parents were not irrevocably merged in their offspring like liquids of different colour but retained their independence and could reappear in subsequent generations. These factors, or genes as we would now call them, segregated with mathematical, and consequently predictable, precision and the characters which appeared in varying proportions were the consequences of their interaction together. I assume that I need not go into the details of Mendel's results and it must therefore suffice to say that his work laid the basis of genetic understanding for all times. Alas, when in 1865 he described his studies to the Brno Natural History Society no one present understood them. Neither did any other biologists who might have read them when published, and so for 35 years they lay unappreciated on library shelves. Following the rediscovery of Mendel's work in 1900, William Bateson, director of the John Innes Horticultural Institution, was Mendel's most prominent protagonist in this country, and it was he who saw to it that his paper was translated into English; the Royal Horticultural Society published it in volume 26 of their Journal in 1901. Horticulturalists evidently played a most important role in bringing the first light of scientific genetics into Europe. But horticulture had another part to play. If we examine the first volume of the *Journal of Genetics,* an English publication meant to display the results of genetical studies, published in 1910, we find that the first paper deals with the inheritance of flower colour in *Primula sinensis,* the second with tuber colour inheritance in potatoes and two further papers with inheritance of doubleness in *Petunia* and Stocks respectively. These papers were a true indication of the fact that subsequent to the rediscovery of Mendel many chose cultivated plants, often ornamentals, as their guinea-pigs for further experimentation. This is not surprising when we remember that for centuries the early hybridizers, as we might call them, had brought wild plants into their gardens and raised them from seed, noting and maintaining variants which were not seen in the wild. They selected and crossed them to produce thousands of garden forms, and added still further to the range of cultivars by bringing together plants from the four corners of the earth and hybridizing them. Such early activities had clothed our gardens in a host of attractive and useful plants and had at the same time exposed wide ranges of genetic variation. Here was the ideal material for the Mendelian geneticists, who must at least have doffed their hats to those who had long before produced it. Horticulture then made the task of the 20th century geneticists easier, and indeed, it was in horticulture that many of the early discoveries were made. But we must not suppose that horticulture has ceased to be of use to genetics, for today the activity of the plant collector, the selectionist and the breeder provides material which still gives a deep insight into the occurrence and location of genetic variation and the way in which it evolves, either in the wild or in cultivation.

The development of genetics

Mention was made of the chromosome in the opening paragraphs of this talk. It did not appear in dealing with the rise of Mendelian genetics because its role in inheritance was not appreciated during this early period. When the importance and full implication of the chromosome did become realized there were many who devoted their time to its study both as the material basis of heredity and as indicators of breeding behaviour, species relationships and pathways of evolution, and so genetics began its bifurcation into experimental breeding and chromosome study respectively. Here we perhaps should realize that knowledge of gene control of characters can only proceed along experimental lines, assessing parents and progeny over several, perhaps many generations. Chromosome constitution and behaviour, in contrast, can be observed directly, without the necessity of breeding, though this is often desirable. Since, therefore, it is possible to observe chromosome number, form and behaviour using simple techniques which may take only a few hours, it may not be surprising that for many species, and indeed for many cultivars, we are in possession of basic chromosome information but not of gene constitution or gene interaction. This situation, of course, does not prevail in the case of crop plants where intense study of both genes and chromosomes is a necessary prerequisite to modern breeding. Here the production of new varieties of tomatoes, cucumbers and lettuce and other horticultural crops must be based on the thorough analysis of these two components of genetic constitution. After all, a desirable variety is one in which the breeder has combined together those genes which will satisfy the grower, transporter, marketer, and consumer. To put them together in this way demands that the effect and inheritance of each has been determined and their several interactions, when placed together in one plant, known. This sometimes also requires that the genetics of some characters of wild relatives are also understood, since it is often the case that these possess one or more genes which can with benefit be incorporated in a new variety. Equally the breeder may have to be familiar with some aspects of the genetics of pathogenicity of viruses or bacteria in order to produce and maintain disease resistant stocks. Modern plant breeding of this sort is a complex and often mathematically based science, and at this time I only mention it in the belief that you are already convinced that genetics is highly relevant here. Ornamentals are another matter for with them the same intensity of genetic effort has rarely been applied, certainly in this country, and we have to rely on new products which often have their origin in America or on the Continent. The amateur, as we know, has made many contributions to our garden varieties and it is he, I think, who may need to be convinced that genetics has a relevance to his activities. It is, therefore, to the amateur that I mainly direct myself this afternoon. That being so, I want now briefly to survey the general utility of genetics and end with an example which will illustrate the use of some of these principles.

The application of genetics

The determination of chromosome number of plant introductions is now a standard procedure in plant breeding institutes and botanic gardens, and is the first step in detecting variation. At the Jodrell Laboratory, for example, we have made such chromosome counts in hundreds of species and thousands of plants—sometimes finding uniformity within a species or genus, but often detecting considerable diversity. I would emphasize here that chromosome differences are rarely suggested in the external morphology of the plant. The result of such studies of this sort are the many chromosome counts listed in published chromosome atlases which present us with a useful survey of the chromosome characteristics of taxa. We find, for example, that woody, long-lived plants tend to be rather uniform in chromosome number, whilst all others will show greater or lesser degrees of variation. *Rhododendron* is a genus which is mostly diploid, but polyploidy can be found, but mostly restricted to south-west China and eastern Nepal. Here too we know that higher polyploids occur at the highest altitudes. In contrast, the genus *Crocus* is chromosomally very complex, with a wide range of chromosome numbers at diploid and polyploid levels. Again we find that the Aloes, Gasterias and Haworthias are relatively uniform whilst Aroids are very variable. And so it goes on—we start to count chromosomes and find situations which we did not expect and we soon realize that chromosome change, either at the numerical level or in their morphology, is a common accompaniment to evolution.

Faced with this fact, and with the information which has been acquired as to the genetic effect of these changes, we are able to explain and understand many phenomena that we encounter in our plants. For example, sterility in most of its forms has a basis in chromosomes or genes. In non-hybrid plants, sterility can be due to uneven polyploidy or to irregular chromosome behaviour at meiosis or to the action of so-called male-sterility genes which inhibit pollen formation. In hybrids, differences in number of chromosomes of the parents may be the cause of sterility, or, in many cases, it is due not to numerical, but structural or genetic difference of the chromosomes. Indeed we can often predict sterility when we know parental constitutions and we can either avoid it, by choosing our parents, or overcome it by using colchicine to double the chromosome number—imitating here the spontaneous doubling which can occur in natural hybrids. Of course when we find gene-controlled male sterility we can make use of it in making hybrids and particularly in the production of F_1 hybrids, as in maize, onions and other crops. Relatively simple investigation is, therefore, an important means of planning our breeding programmes and of facilitating the production of satisfactory cultivars. Of course there are many horticultural plants which are either sterile or, if fertile, do not breed true, and these we propagate vegetatively. This is not always the best means of propagation as it is slow and laborious. Such plants have evidently not been well bred—are often spontaneous mutants which would segregate if grown from seed. We are familiar with the nature of mutation, and in many cases careful breeding based on genetical knowledge could result in true breeding strains. So often, however, the effort involved has not been considered worthwhile.

The impact of genetics on cultivar production can indeed be to maintain purity in our stocks. It is known for example that inbreeding can produce pure lines which will breed true for almost all characters, and equally it is possible to produce homozygosity only for certain specified characters, and these too will be pure breeding. It is possible, again, to produce haploid plants with desirable genes and then colchicine these to give rise to absolutely pure breeding diploids. Haploids can occur normally and may even be unconsciously selected as cultivars, as in the case of the *Pelargonium* variety 'Kleine Liebling', but modern experiments have shown that they can be produced directly from haploid pollen grains, and it is quite possible that haploid production will play an important part in future plant breeding. Again, even heterozygous plants can breed true when they produce seed without sexual fusion (=apomixis). This is a natural phenomenon characteristic of many plants such as *Rosa canina, Taraxacum* and others, but again genetical studies may eventually result in precise knowledge of the biochemistry of the phenomenon and it may be possible to induce it at will. Whether we talk of breeding systems, bud mutations, chimeras, or incompatibility in fruit trees, all have a genetical basis which we can understand and often manipulate to our great advantage. In the production of food, fibre and oil plants, genetics has been found to be indispensable. In horticulture it has found its place in some areas, but has been sadly neglected in others. The early genetic interest in ornamentals was not sustained, but undoubtedly it will eventually come to play its rightful role.

Finally then, let me quote a practical example of the use of genetical observations in the production of a novel type of plant— *Delphinium* with red and orange flowers, as this I think may provide a useful demonstration of the use of several of the features of genetical study which I have so far mentioned.

The breeding of red Delphiniums

You will I am sure be aware of the success which Dr Legro of Wageningen has had in the production of red Delphiniums. His experiments and breeding procedures have of necessity been based on a sound understanding of genes and chromosomes. It is instructive to see that although red and orange species were known to exist in the genus during the last century, that it so say, *Delphinium cardinale* and *D. nudicaule* respectively, attempts to introduce their flower colour genes into cultivars of *Delphinium elatum* failed because the cross between them would not succeed. Modern chromosomes counts show the wild species to be diploids and *D. elatum* tetraploid. This seemed to be a likely basis for the inability of the species to hybridize. By applying colchicine Dr Legro was able to produce tetraploid forms of both *D. cardinale* and *D. nudicaule,* but, although this did allow hybrids to be made with *D. elatum,* the progeny were not of desirable types. So he decided to hybridize the two diploid species, but now their F_1 was sterile because of the differences in the genetical structure of their chromosomes. Colchicine was again applied, but this time to their F_1 hybrid which resulted in the production of a fertile tetraploid hybrid, and this, when crossed to *D. elatum,* gave rise to progeny showing the potential for the production of good red and orange cultivars. Here then we see procedures based on the understanding of chromosome differences, on the appreciation of the reasons for hybrid sterility, and of the usefulness of colchicine. As a result we are now able to see successful red and orange Delphiniums. This is really an excellent example of the relevance of genetics to horticulture.

DISCUSSION

Dr JONES, in response to a request to describe very briefly his method of determining the chromosomes of a plant, said that the method was essentially simple. One took a young actively growing root tip and one physically squashed it, and then one placed the squashed tip in stain and looked at it under the microscope. However, it was a little more complicated than that, for one had to treat the root, quite simply, with some chemicals, but after that one could examine the chromosomes without too much difficulty. He said that he had, in his laboratory, young girls who had been trained in the techniques, and after three months were highly efficient and did chromosome counting as a purely routine process. It was not a difficult thing once you knew how.

Mrs K. N. DRYDEN asked why, when an amateur sent a plant to a botanic garden to have its chromosomes counted, all they received were letters saying that the plant had been killed before the chromosome number had been obtained.

Dr JONES replied by asking which botanic garden and which plant?

Mrs DRYDEN said that the plant was a *Lewisia* and there were two botanic gardens involved, one in England and one in America.

Dr JONES replied by saying that if Mr Ingwersen was still present he thought he would say that Kew Gardens did in fact count the *Lewisia* for him. But he believed the Americans had failed. Counting chromosomes was a relatively simple procedure, but it did require a fair amount of man power and time. It was not surprising really that botanic gardens did not offer a chromosome counting service. One could well imagine what would happen if that occurred. Dr Jones was sure there were many who would love to know the chromosome numbers of their plants, but that was physically impossible.

Mr C. D. SAYERS, remembering that it had been said that chromosome numbers increase with altitude in the Eastern Himalayas but not in New Guinea, asked if there was any reason why this phenomenon should not occur in the New Guinea mountains?

Dr JONES replied that although the occurrence of polyploidy in the uplands of SW. China and Nepal was a practical occurrence, he could not explain why it should occur there and not in the mountains of New Guinea. In fact, the reason why he looked at New Guinea plants was the thought that at these altitudes of 9,000 ft. or so one might well find polyploids among them, but this was not so. It was possible, of course, that the New Guinea plants were of more recent origin than those in India or that the opportunities for hybridization, which frequently preceded polyploidy had not been present in New Guinea, in contrast to the Himalayas, but further than that he could make no suggestion.

Dr W. T. STEARN commented that he had seen Himalayan Rhododendrons growing wild. He had also seen upland Rhododendrons in their native habitat in New Guinea, and would say that the Himalayan ones grew under much tougher conditions. Because of this he felt that selection pressure was probably much higher and more severe in West China and the Himalayas than in New Guinea.

Dr JONES replied that if the conditions were widely different then this could be accepted as a very good explanation.

HYPERICUM

N. K. B. ROBSON

Department of Botany, British Museum (Natural History)

Introduction

   Hypericum has been known as a garden genus for a long time yet it continues to play its part in the kaleidoscope of colour and form that is the modern garden. It has one drawback, however, that should be mentioned right at the beginning: the flowers of all but a very few of its 400 or so species are yellow. But, having said that, it must be pointed out that the shades of yellow range from pale lemon to apricot, and are frequently relieved by tinges of red, especially in bud.

   In contrast to the relative uniformity of flower colour, *Hypericum* shows an extremely wide range of variation in many other characters—in habit, for example, from trees up to 10 m in height in the East African mountains to tiny annuals not much more than 2 cm high. Even the predominant capsule has given rise in four quite separate parts of the genus to a berry or berry-like fruit.

   Although most of the best-known garden Hypericums are exotic in origin, such as the widespread *Hypericum* 'Hidcote', the 16 native or naturalized species include four that are common (I might almost say 'standard') garden plants, namely *H. calycinum, H. hircinum, H.× inodorum* and *H. androsaemum*. These are all shrubs; but I hope to show that the herbaceous species are, or can be, of horticultural value too.

The shrubby species

   The shrubs that I have just mentioned differ from the remaining herbaceous species in having five separate bundles (or fascicles) of stamens which, along with the petals, fall before the fruit develops. Of these, the most useful must surely be *H. calycinum* L., the Rose of Sharon or Aaron's Beard. It is a native of north-western Turkey and adjacent Bulgaria, where it forms part of the ground flora of dry or damp deciduous woodland and also occurs on shaded banks. It was introduced to Britain from near Istanbul in 1676 by Sir George Wheeler, at that time the British ambassador to Turkey, and soon became established in British gardens. Indeed, it has become so popular that one tends to take it somewhat for granted, forgetting how useful it is in that most difficult of garden habitats— dry shade. It is also widely and effectively used to cover slopes, particularly shaded ones where little else will grow well. Of course, it will flower most freely in sunny situations. *H. calycinum* has a reputation for being invasive, spreading by means of shallow rhizomes, but it is nevertheless an invaluable plant in the right site. It must have achieved its present widespread distribution in semi-natural habitats in the British Isles largely as a result of vegetative spread, because our autumns are usually too damp to allow seed to ripen in the capsules.

The remaining shrubby species are all members of section *Androsaemum* and are closely related to one another: they include the only true native, *H. androsaemum* L. This is a plant of woodland margins and lightly shaded, not too dry habitats, and occurs in several different forms. These are not all equally effective in gardens, as some have small, rather insignificant flowers and green leaves and sepals, whereas others with larger flowers tend to have leaves and sepals tinged red with the sap that gives the species its name—*androsaemum* ("Man's Blood"). The variation in *H. androsaemum* extends to the shape of the ovary, the length of the styles and, more important horticulturally, the fruit. This is always fleshy; but in the larger-flowered red-tinged forms it tends to be larger and ovoid, to redden early and to remain dark reddish brown. On the other hand the smaller-flowered forms produce smaller, cylindrical-ellipsoid fruits that ripen later to a brighter red before becoming shiny black. These fleshy fruits are very attractive to birds, and by the late autumn there are rarely any left on the plant. It is therefore difficult to be sure whether they would persist if left alone or whether they are in fact deciduous, as indicated in Floras. At any rate, birds do distribute the seed of this species, so that seedlings frequently appear even in gardens where no plants of it have been growing. As a result of this method of dispersal and its ability to flourish in gardens and semi-natural habitats far to the north-east of its natural distribution in Britain, there is an area in north-east England where it is difficult to decide whether or not it is a native.

Hypericum androsaemum has a close relative in *H. hircinum* L., which is equally variable but can always be distinguished from it by several characters, for example the longer, straight (not curved) styles, smaller deciduous sepals, tardily dehiscent capsular fruits and the presence of caprylic acid, giving it the goat-like smell from which it gets its name. Although not a native, it was introduced before 1640 into our gardens from the Mediterranean region, where it has a discontinuous distribution. Over this region it varies very much in size, from a tall shrub up to 1·5 m high, with large acute leaves and large flowers (in the Levant) to dwarf shrublets of 30–60 cm with small acute leaves and small flowers (in the Balearic Islands). This is the form known in gardens as var. *pumilum* Watson. In addition, there are smallish forms on Corsica, Sardinia and Crete with rounded leaf apices. In the British Isles, *H. hircinum* has escaped from cultivation sufficiently frequently to have gained a place in British Floras, but it has not become so widely established as

H. calycinum. It has, however, been involved in the production of another plant of our gardens which, in its turn, has become naturalized.

I refer, of course, to *H. inodorum* Miller, which is still better known by the name *H. elatum* Aiton. This plant was first described in 1768 by Philip Miller from a specimen growing in Chelsea Physic Garden, although Aiton says that it was first recorded at Kew in 1762. As it looked rather like *H. hircinum* but did not have the characteristic smell of that species, Miller gave it the epithet *inodorum,* just as Willdenow did some years later when describing a Caucasian species that reminded him of *H. hircinum* but lacked the characteristic smell. This second '*inodorum*' is now known as *H. xylosteifolium* (Spach) N. Robson.

When the characters of *H. inodorum* are examined, all those in which it differs from *H. hircinum* (for example larger, persistent sepals, shorter styles, somewhat fleshier fruit and rounded leaf apices) tend in the direction of *H. androsaemum,* which suggests that it may be the hybrid *H. hircinum* × *androsaemum.* Supporting evidence for this theory is provided by the field records for intermediate plants of this description from natural habitats, which are from the areas in the western Mediterranean and northern Spain where *H. androsaemum* and *H. hircinum* occur together as wild plants. I have not yet tested the hybrid theory by crossing the suspected parents but a plant reported to be of this parentage, which was sent to me by Mr Graham Thomas, proved to be *H.* × *inodorum.*

As we have already seen, both *H. androsaemum* and *H. hircinum* are highly variable species, and so it is not surprising that *H.* × *inodorum* shows great variation in habit, in size of leaves and flowers, in size, colour and fleshiness of fruit, and so on. One form that has become popular in gardens recently, *H.* × *inodorum* 'Elstead', has bright rose-red, semi-fleshy fruits, as well as other characters which suggest that it may be the result of a back-cross of *H.* × *inodorum* to *H. androsaemum.*

Hypericum × *inodorum,* then, is an addition to the British Flora that appears to have originated in our gardens.

The herbaceous species

Although I have no time to discuss the herbaceous species in detail, I should like to point out one or two that can be effective in gardens. *H. perforatum* L. and its allies are usually regarded as being too 'weedy'; but I know of a dense stand of *H. maculatum* Crantz subsp. *maculatum* (the rarer one in Britain) that forms an attractive feature in the rock garden at the Cruikshank Botanic Garden, Aberdeen, whilst the bright-red tipped-buds and wavy-margined leaves of *H. undulatum* Schousboe ex Willd. would look well in those damp spots where Primulas, Astilbes and Hostas flourish. *H. linarifolium* Vahl, which in the wild grows on dry sunny banks or cliffs, can also look well in a rock garden, as I have seen in the Royal Botanic Garden, Edinburgh. Its close relative, *H. humifusum* L., also prefers open situations and, being a shallow rooter, can grow happily where there is very little soil. Although individuals are short-lived, it is a charming little plant which will seed itself in such places as paving cracks and path sides. Finally, I must mention *H. elodes* L., which can be effective in the mud and shallow water at the margin of a pond or lake if the conditions are not alkaline. It seems to be difficult to establish, in my experience, and the one cultivated example that I know of died after about seven years' cultivation; but I think that it could well be grown successfully in suitable conditions. Quite apart from its considerable botanical interest, its attractive foliage and unusual form could make it a worthwhile addition to the bog garden.

DISCUSSION

Mr J. S. L. GILMOUR asked why there had been no mention of what he thought was one of the most beautiful of British wild flowers, *Hypericum pulchrum*.

Dr ROBSON replied that his only plea was insufficient time. He did agree that the species well deserved its epithet, *pulchrum*.

'MESEMBRYANTHEMUMS'
ESTABLISHED IN THE ISLES OF SCILLY

J. E. LOUSLEY

The Isles of Scilly have exceptional advantages in their winter climate. Severe frosts are rare and seldom damage plants in sheltered places. A wide range of the more showy frost-tender *Aizoaceae* have been grown at Tresco Abbey where 50 "species" were in cultivation by 1852 and by 1947 the garden list claimed 153. Small scraps stuck in walls and rocks grow very easily and in this way were spread to gardens, lanes and cliffs on all the inhabited islands.

Of all the "mesems" grown, only one, *Carpobrotus edulis,* deserves serious consideration as a British plant. This is abundant on the cliffs of Devon and Cornwall as well as in Scilly, competes successfully with native plants, and has ecological significance. It is spread by gulls as well as humans and has persisted for a long time. Nevertheless it is very sensitive to frost and in a severe winter is killed in exposed places, even in Scilly. Climate restricts its distribution in Britain and it is still possible that an exceptionally hard winter would remove it from our list. Other species of *Carpobrotus* are found in Scilly, SW. England and the Channel Isles.

Erepsia heteropetala is one of the very few members of this group which seems able to spread from seed in Britain. This unattractive plant grows on the face of a long disused quarry in St Mary's and I suspect that it was thrown out by a gardener in disgust. It fruits abundantly but the flowers are small and inconspicuous, though the leaves sometimes turn a rich red.

The remaining species grow where they have been planted even though at times they may seem impressively "wild". Thus *Drosanthemum floribundum* (Plate I) with its mass of attractive Michaelmas Daisy-like flowers, and its leaves glistening with minute crystalline cells, is common on banks near cottages. On The Garrison headland it has been planted to cover the sewage outfall and has spread vegetatively to the adjacent rocks. Similarly, *Disphyma crassifolium* drapes its chains of fascicled cylindrical leaves down the low cliffs at New Grimsby on Tresco. My enquiries revealed that it may have been there for nearly a century, fortunate in an exceptionally frost-free habitat and with very little competition from native species. It behaves similarly on the mainland on bare cliffs by the sea.

Oscularia deltoides (Plate I), with its curious deltoid leaves, is seldom far from houses, except when deliberately planted on stone walls round bulbfields. Similarly, *Aptenia cordifolia*, with flattish

FIGURE 1

1. *Oscularia deltoides* (leaf and leaf section, ×1). 2. *Erepsia heteropetala* (leaf, ×1). 3. *Disphyma crassifolium* (leaf, ×1½). 4. *Carpobrotus acinaciformis* (leaf section, ×½). 5. *Drosanthemum floribundum* (leaf, ×1½). Reproduced by kind permission from J. E. Lousley, *Flora of the Isles of Scilly* (David & Charles).

cordate leaves, which has found a niche in many parts of the world with milder winters than ours, shows no tendency to spread by natural means, but persists for a time where planted. They owe their persistence to an ability to grow, until killed by an exceptionally cold winter, on bare places unattractive to native plants.

Ruschia caroli stands up to competition a little better. It is winter flowering, and its more erect woody stems take it up to a level where it can withstand limited smothering by blackberries and gorse. It may be also that its competitors shelter the lower parts from frost and thus help longer term persistence.

I have had time to mention only a few of the species in Scilly, and there are others on sea cliffs and on quarry faces elsewhere in Britain, or which have a fleeting appearance on refuse tips. These plants cannot be completely ignored by people interested in the British flora and I am collecting information for a full account. Nevertheless, they are of insignificant importance to field botanists compared with that of fully hardy groups established in wild situations. A great deal of information about the horticultural history and garden experience of such plants as the Asters and Cotoneasters is essential before satisfactory botanical accounts can be prepared. It is in such groups that there is a profitable field for collaboration rather than in plants dependent on a run of mild winters and deliberate planting.

THE PRINCIPLES OF BOTANICAL NOMENCLATURE, THEIR BASIS AND HISTORY

WILLIAM T. STEARN

Department of Botany, British Museum (Natural History)

Introduction

The *International Code of Botanical Nomenclature,* which regulates the scientific naming of plants, is a long and elaborate document resulting from earnest consideration and sometimes bitter debate by experienced taxonomic botanists going back to 1867 and it embodies traditional procedures established even before then. The present edition of the *International Code,* which the Eleventh International Botanical Congress adopted at Seattle in 1969, was published at Utrecht in 1972. It begins, as did earlier editions, with six principles underlying its rules and recommendations and all these follow from the general statement in its preamble: 'Botany requires a precise and simple system of nomenclature used by botanists in all countries....... The purpose of giving a name to a taxonomic group is not to indicate its characters or history but to supply a means of referring to it and to indicate its taxonomic rank. This Code aims at the provision of a stable method of naming taxonomic groups, avoiding and rejecting the use of names which may cause error or ambiguity or throw science into confusion. Next in importance is the avoidance of the useless creating of names'. A principle is a fundamental truth, or at least something accepted as such, which serves as the basis of reasoning and procedure. My intent here is to indicate the background to the principles of the *International Code* chiefly by means of details exemplifying their history and thereby to demonstrate their aims and utility.

The basic purpose of botanical and horticultural nomenclature is that of all human communication, namely, mutual intelligibility in a manner precise enough to meet the needs of the occasion. This, of course, can only be achieved by ensuring that a given combination of sounds or letters conveys to all concerned the same meaning or designates the same object. In other words, the giver and the receiver must have agreed that for both of them this combination is to be associated with a particular meaning; without such linguistic common ground there can be no satisfactory mental ebb and flow between them. Its development is naturally gradual. A small child in an English-speaking community, for example, learns from his parents and teachers that the word *knife* means something on the table with which he can cut and hurt himself unless careful in its use; it refers to a concrete object. He learns next that the word *knife* refers to one class of eating instrument, *fork* and *spoon* to other classes, and so on.   In due course he learns further that the essential features of a knife are a handle for holding and a blade for cutting and that, although such species of the genus *knife* as the *bread knife* the *carving knife*, the *table knife* and the *steak knife* are used in the kitchen and on the meal table, other species such as the *paper knife*, the *pocket knife* and the *pruning knife* have special non-culinary uses connected with paper, pencils, shrubs, etc.   The word *knife* is, for him, now a generic name.   This development of application exemplifies the capacity for generalization, of forming an abstract concept out of varied concrete examples by recognizing their common feature, which underlies all biological classification.

Thus early in life the child becomes a taxonomist at the breakfast table.   Without realizing it he becomes acquainted not only with the concept of *genus* in its logical sense, i.e. a class of objects which can be divided, and of *species*, the objects into which that class is divided, together the *summum genus* and the *infimae species*, but also with binomial nomenclature for the species.   Such commonly used specific designations as *bread knife, carving knife* etc. are always two-word names, just as are the scientific names of biological species like *Alnus glutinosa, Alnus incana, Alnus orientalis* etc. This two-word or binomial system of naming kinds within a group, whether plants, animals or artifacts, is common to almost all languages, hence it must go back to a very early stage in the development of human conceptual thought and speech when our very remote ancestors began to distinguish between the general and the particular.   It was not invented by Linnaeus in 1753, as sometimes stated!   Essentially it consists of using one word, such as *knife* or *Alnus,* for the genus or group as a whole and another word qualifying this to indicate the particular species or kind, such as *bread, carving, paper, steak* or *glutinosa, incana, orientalis* and so on. Such a name as *bread knife* or *Alnus orientalis* is an agreed designation for an object, not a description of it.   In botany the use of the binomial system has become standardized and specialized. Thus, for the species named *Alnus glutinosa* and *Alnus orientalis,* the generic name *Alnus* applied to both implies that they both resemble each other in possessing the characters which distinguish the genus *Alnus* or alder from other genera such as *Betula, Carpinus, Quercus* etc., but the specific epithets *glutinosa* and *orientalis* imply that they nevertheless possess their own individual characteristics. Words such as *knife, carving,* and so on have meaning only within a particular language community, here the English-speaking one. In the German-language community the child would learn the word *Messer,* in the French-language community the word *couteau.* Mutual intelligibility between different language communities is achieved by the recognition of equivalents, e.g. *knife=Messer= couteau,* each of which can be associated with the same concept. There is now no one internationally accepted non-national word for such a concept as 'knife'. The Latin *culter*, except maybe in a monastery, is no use at all for the purpose. Botanists and gardeners are thus fortunate in forming an international language community wherein such a name as *Alnus* has an accepted application all the world over; it is a name to which the more restricted names of different language communities, e.g. English *alder* or *aller*, French *aune*, German *Erle*, Dutch *els*, Danish *el*, Norwegian *or*, Swedish *al*, Italian *ontàno*, Spanish *aliso*, Polish *olsza*, Russian *ol'kha*, Roumanian *anin*, can be made equivalent. The development of this Botanical-Latin-language community in modern times is a belated legacy of the former political dominance of the Romans and the later cultural dominance of their language in Europe. A little knowledge of both makes understandable some provisions of the present *International Code of Botanical Nomenclature*.

The use of Latin

During the early, formative stages of language, when human communities were necessarily small, agreement that a particular combination of sounds should indicate a particular object was doubtlessly made by a few people, simply following the usage of a leader in the community, and it then passed from the speech of the family into the speech of the tribe and in the fullness of time maybe into that of a nation or even an empire and thence into a widespread language community, although liable to distortion and change on the way. Some plant names such as *Rosa, Lilium, Hyacinthus, Crocus* and *Zingiber* had their origin in very remote antiquity, probably long before the Latin language took the shape we know. This began as the speech of the tribe Latini inhabiting the land around Rome. It developed into the major language of the Italian peninsula after the Latini, centred on Rome, had triumphed militarily over their neighbours, the Volscians, the Sabines, the Etruscans, and so on. The conquests of the Romans made Latin the common language of an empire extending from north Africa northward to Britain and eastward to western Asia. It encased relics of supplanted Mediterranean languages and it was enriched by borrowing concepts and words from the Greeks. The Romans thus created an international language community, embracing peoples originally with very different languages of their own. The later spread of Christianity over the Roman empire and beyond its bounds made Latin the official language of the Christian Church based on Rome; as the English translators of the King James version of the Bible said, 'this tongue also was very fit to convey the Law and the Gospel by, because in those times very many countries of the West, yea of the South, East and North, spake or understood Latin, being made Provinces to the Romans'. Out of the subsequent use of Latin during the Middle Ages and later, for ecclesiastical, diplomatic, legal and cultural purposes, for hymns and poems as well as for Papal bulls, has come our use of Latin today as the basis of our botanical nomenclature.

Those who spoke and wrote Latin in the Middle Ages did not bother their heads overmuch about linguistic purity, so long as they could understand one another, any more than did the Roman themselves, who took lots of words from Greek, among them many plant names such as *Achillea, Anemone, Daphne* and *Scilla*. Modern botanists follow the medieval habit of coining new words and adapting old words to meet current needs. Latin thus survived as an adaptable independent standard language, belonging to no one national group, long after it had ceased in its original widespread form to be the vernacular tongue of any one people; its descendants through vulgar Latin diverged into Italian, Spanish, French, Roumanian and other languages, each becoming more and more unintelligible to outsiders. Consequently, when even in Europe most people were illiterate, Latin down to the 16th century probably had more readers and was more widely understood than any one national language. In the 16th and 17th centuries any scholar who wished his work to be widely read naturally wrote it in Latin, 'the paper language of scholars'. Hence the herbalists of the 16th century used Latin names for their plants and cited vernacular names in English, German, etc. as synonyms. Thus they formed, in Karl Vossler's words, 'an empirical language community . . . held together by the will to work at a common language material as the special instrument of mutual understanding'. Their use of Latin in this way during the 16th century, when botany developed out of pharmacy, was of crucial importance. Had the vernacular languages of Europe been then so well developed for technical use and the number of interested and literate men in every country been big enough to bear the cost of printing learned works in their own vernacular language, then Latin might have been lost as a means of international communication before botany had become a science needing it; in consequence there might not now be an international system of recording information under botanical names conforming to the *International Code*. These 16th-century herbalists, Brunfels, Fuchs, Ruel, Gessner, Turner, Mattioli, Cordus, Dodoens, L'Obel, L'Ecluse, Cesalpino, Colonna, Gaspard Bauhin and others, by their use of Latin alongside their vernacular languages, thereby established a tradition and a procedure which Linnaeus inherited in the 18th century and which has proved of lasting value. In consequence it is a principle (no. V) of the *International Code* that '*Scientific names of taxonomic groups are treated as Latin regardless of their derivation*'.

Correct application of names

There were, however, many difficulties in the 16th century as well as later over the right application of the Latin names of plants. Mutual intelligibility vanishes if people apply the same name to different things. They must apply it to the same object and have some means not only of keeping it so applied but also of ascertaining to what object a particular name applies. The provisions of the *International Code* relating to priority and typification are intended to bring about this stable application of names, even though they sometimes appear to result in anything but that!

Two small books illustrate the nomenclatural difficulties of the 16th-century pioneers, namely, the *Botanicologicon* (1534) by Euricius Cordus (1486–1535) the German classicist, poet and physician and the *Libellus de Re herbaria* (1538) by William Turner (c. 1508–1568) the English divine, naturalist and physician. Both of these doughty Protestant controversialists were concerned with the correct naming of the herbs used in medicine; they wished to use correctly the Greek and Latin plant names received from the Ancients, notably Dioscorides and Pliny. The *Botanologicon* takes the form of a long discussion about plants and their names between Cordus and his friends Johannes Meckbach, Anton Niger, Guillaume Bigot and Johannes Ralla. Cordus bemoans that most pharmacists, indeed almost all, are imprisoned in such ignorance that they cannot recognize half the herbs prescribed by physicians. He and Turner attempted first to correct and standardize usage by means of equivalents, i.e. what is called *x* in one language is called *y* in another, as illustrated by the *knife=Messer=couteau* example above. Thus Cordus says that he believed *Cicerbita* to be the plant which Germans in Saxony called 'Hasenkohl' (i.e. *brassica leporina*) and in Hessen 'Gänsedistel' (i.e. *carduus anserinus*) but that others believed *Taraxacon* to be identical with this; it was called *Sonchus* in Greek. Turner says 'Cicerbita a grecis sonchos, a nostris Sowthystell appellatur', i.e. 'sow thistle' by us. From this one could conclude that the English name 'sowthistle' and the German 'Hasenkohl' referred to the same plant for which the learned name could be either *Cicerbita* or *Sonchus*.

The major difficulties of these early herbalists came from the lack of agreed and hence fixed reference points. Thus Cordus and his friends argue whether the names *Lingua cervina, Scolopendrium, Asplenum* and *Phyllitis* refer to the same plant. In the end Cordus says that 'Scolopendrium is not Lingua cervina as almost all physicians believe but the plant that the Latins called Calcifraga and today in Arabic is the Ceterach of pharmacies, in German Steinfarn, Spikant' and that 'Phyllitis is really Lingua Cervina, a plant other than Scolopendrium'. Turner in 1547 says 'Asplenum or asplenium named in greke asplenon, or Scolopendrion, in duche Steinferne, is called of the poticaries Citterache'. The plant he had in mind is the rusty-back fern, *Ceterach officinarum* DC. (*Asplenium ceterach* L.) Turner also says 'Phyllitis as Cordus judgeth, is the herbe which we cal in englishe Hartes tonge, the duch cal Hirtzen zumgen, the french men Lang de Cerfe, the poticaries Linguam cervinam. To whose judgement I rather assent, than to Ruellius and Fuchsius'. The matter was evidently confused and controversial. The plant Turner had in mind is the hart's tongue fern, *Phyllitis scolopendrium* (L.) Newman (*Asplenium scolopendrium*

L., *Scolopendrium vulgare* Smith). Thus here we have Brunfels using the name *Scolopendrium* for the hart's-tongue fern but Cordus stating that this belongs to the rusty-back fern. Gaspard Bauhin in 1263 got round the difficulty here by putting the first under the heading *Phyllitis*, the species being named *Lingua cervina officinarum*, and the second under the heading *Asplenium sive Scolopendria*, the species being named *Ceterach officinarum*. Linnaeus put both species in the genus *Asplenium* and named the first *A. scolopendrium*, the second *A. ceterach*.

Illustrations and Herbarium Specimens as Fixed Points of reference

The standardization of names only became practicable when a quantity of good illustrations had been published; these provided checks as to identity and so made stable the use of names hitherto passed on only by word of mouth from master apothecary or physician to apprentice or student and thus liable to varying application. The first of such illustrations drawn from nature and cut for printing from wooden blocks were those in Brunfels, *Herbarum vivae Eicones* (1530–1536), and Fuchs, *De Historia Stirpium* (1542), dealing with wild and cultivated plants mostly of the Rhineland region. They were followed by the illustrated works of Mattioli, Dodoens, L'Obel, L'Ecluse (Clusius) and others. Thereafter anyone wishing to know, for example, what kind of plant *Campanula vulgatior foliis urticae* C. Bauhin (*Campanula trachelium* L.) is, could be referred to the woodcut in Fuchs, *De Historia Stirpium* 432 (1542) as *Campanula*. Even such fixed points of reference did not wholly remove the possibility of misapplication of names since the illustration and accompanying text might fail to give the information necessary to distinguish between closely allied species.

Early, however, in the 16th century an Italian professor at Bologna, Luca Ghini (c. 1490–1556), had discovered that plants pressed and dried and then gummed on to sheets of paper could be used like illustrations and sent to correspondents, among them Mattioli, as permanent records. Such dried specimens, together forming a *hortus siccus* or 'dry garden', were at first mounted in books, a practice which continued well into the 18th century. This prevented the loss of specimens, but paralysed their classification; they had to stay where they had originally been stuck. Linnaeus, however, in the 18th century mounted his specimens on loose separate sheets of paper. He could re-arrange these as his ideas on classification changed, one specimen could be readily compared with another by placing them side by side, new specimens could be placed next to the old ones they most resembled. The fixed *hortus siccus* developed into the adjustable herbarium able to reflect accepted classification and promote research. Its specimens could now serve as fixed points of reference for names, being permanently available for re-study and hence potentially of much greater scientific value than illustrations which are necessarily limited

by the knowledge of their period as to the details they portray.
Thus it has been possible to deduce the chromosome number of
some specimens in the Linnaean Herbarium by measuring their
pollen grains, thereby obtaining from them information which no
contemporary illustration or description could have recorded.

Pre-Linnaean names

This provision of illustrations still left unsettled the choice
of names for the plants themselves. In the 16th and 17th centuries
there existed no generally accepted system of classification and
no rules of nomenclature. Hence people named plants as they
pleased. There was no such general stability of nomenclature as
the *International Code* has promoted. The name *Hesperis matronalis*,
bestowed on that old garden plant, the dame's violet, by Linnaeus in
1753, has remained its correct name since then; it has been stable
for more than two hundred years. By 1623, when Gaspard
Bauhin attempted to standardize names, this had acquired, according
to his entry, the following names:
' I. Hesperis hortensis
   Viola alba, persica Hermolai, Trag.
   Leucoium & Viola purpurea, Fuch.
   Leucoium Diosc. album, purpureum, Fuch. ico.
   Viola alba, Tur.
   Viola matronalis, Gesn. hor. Dod.
   Viola purpurea & alba, Lugd.
   Viola matronalis sive Damascena, Ad. Lob.
   Viola sylv. alba & purpurea, Caes.
   Hesperis 3, Clus. hist. ico. nostra, Cam.
   Viola Damascena flore purpureo, Svvert. '
That makes eleven names of varying length for this species within
less than a hundred years, almost all of them equally acceptable
then in the absence of any agreed system of nomenclature. To
remove such ambiguity, principle IV of the *International Code*
reads '*Each taxonomic group with a particular circumscription,
position and rank can bear only one correct name, the earliest that
is in accordance with the Rules, except in specified cases*'. By the end
of the 17th century the situation as regards names had become
much worse because the number of plants known had much increased
ans so too had the difficulty of recognizing them. Floras of the
modern type with keys and convenient, methodically drafted
descriptions under standard names had not then been conceived,
let alone written. Moreover, the efforts of botanists to make their
names as informative as possible inevitably lengthened these
inordinately. Thus Plukenet in 1696 used the name *Clinopodium
angustifolium non ramosum, flore caeruleo: labio trifidis atropurpureis
maculis ornato* for the species later named *Monarda ciliata*.
Dillenius in 1732 named *Verbena bonariensis altissima, lavendulae
canariensis folio, spica lavendulae* the species later named simply
V. bonariensis. Hundreds of such long-discarded many-worded

descriptive names appear in the synonymy of species given binomials in Linnaeus's *Species Plantarum* (1753).

Linnaeus's reform of nomenclature

The change from 17th-century to modern nomenclature was primarily due to the Swedish naturalist Carl Linnaeus (1707–1778). Linnaeus was born in humble circumstances, the son of a country parson in a war-impoverished country. He had to make his way in the world and up the academic ladder to the prestige of an Uppsala professorship entirely by his ability, his industry and his flair for making his merits and achievements known to people who mattered. One of his opponents, Casimir Friedrich Medicus, indeed remarked that 'A spirit of reform was his total basic purpose and a high grade of conceit his daily companion'. Reform, however, was much needed when Linnaeus was a young man. He dealt first with the genera, providing new definitions and over-often new names for these in his *Genera Plantarum* (1737), and setting out his principles and procedure as regards nomenclature in his *Critica botanica* (1737). To our thinking, his rules were mostly very sensible. He disliked very long names and in consequence he altered names adopted by his predecessors or contemporaries, never a welcome act! One has only to reflect how annoyed present-day gardeners would be, if after having spent time learning how to say *Conocarpodendron, Helminthotheca, Eupatoriophalacron* and *Christophoriana*, they were then told that these should be called *Protea, Picris, Verbesina* and *Actaea*!

Nevertheless, Linnaeus's views prevailed in the long run, because he used such names in works which were much more useful at the time than those of any other author. Thus, in his *Genera Plantarum* his names were associated with detailed descriptions methodically drafted on the same plan covering all known genera. Most of the generic names he used then we still use today.

The genera having been established, Linnaeus turned his attention to their species. His aim was 'to give distinguishing characters which will alone allow the species to be recognized at first glance with far greater certainty than all the descriptions and pictures provide together' (*Critica bot.* no. 256). 'The specific name', he said, 'should distinguish the plant from all others of the genus' (*Critica bot.* no. 257). 'The generic name should be attached to every species of the genus' (*Critica bot.* no. 284). By 'specific name' (*nomen specificum*) Linnaeus did not, however, mean such a name as *Hesperis matronalis* but a diagnostic phrase-name such as *Hesperis caule simplici erecto, foliis ovato-lanceolatis denticulatis, petalis mucrone emarginatis* which he used in his *Hortus Cliffortianus* (1738). Such a name stated the characters by which this species could be distinguished from another species named *Hesperis caule ramosissimo, foliis lanceolatis saepius dentatis*. Although impossible to remember for long and equally impossible to write on a garden label, the diagnostic name was nevertheless regarded by Linnaeus and his followers as the true name of the species, the *nomen specificum legitimum*, because it was a diagnostic name. The function of a name is, however, to be a designation, not a diagnosis. Binomial names like *bread knife, carving knife, pruning knife* etc. by themselves do not enable the objects concerned to be diagnosed with precision, but they serve adequately for general purposes because they can be associated with accepted concepts recorded, for example, in a cutler's catalogue. These long names, usually called polynomials, became less useful as designations the more elaborate they became as diagnoses: these were indeed contradictory functions. Linnaeus solved the difficulty in 1753 by separating these functions, in other words, by giving most species two names. One, a diagnostic or polynomial phrase-name, as above, was for use in books and the other, a binomial or two-word name, such as *Hesperis matronalis* or *H. indica,* was for everyday use.

This two-word specific name consisting of a generic name, e.g. *Hesperis,* plus a specific epithet, e.g. *matronalis,* is not, however, a *proper name* in the sense of formal logic; as defined by J. M. Keynes (1906), 'a proper name is a name assigned as a mark to distinguish an individual person or thing from others without implying in its signification the possession by the individual in question of any specific attributes'. The scientific binomial has, to the initiated, a high information content, especially when associated with the name of the person or persons responsible for its publication. In the binomial *Hesperis matronalis* L., the specific epithet *matronalis,* 'pertaining to a married woman', admittedly gives no clue by itself to the nature of the plant. Use of the generic name *Hesperis* L. implies, however, that the plant has the characteristics of the group of plants to which the name *Hesperis* has been assigned, i.e. it belongs to the botanical family *Cruciferae* and has the technical characters of the stigma which distinguish *Hesperis* from *Matthiola, Malcolmia, Cheiranthus,* etc. and of the pod and seeds which distinguish them from other genera of *Cruciferae.* The epithet *matronalis* by itself merely serves to indicate that species is not the same, for example, as *H. tristis, H. laciniata* etc. and if one turns to a botanical work, e.g. the *Flora Europaea* 1: 275–276 (1964), one can ascertain what its specific features are. The authority 'L.' for Linnaeus indicates much of its history, e.g. that the plant was known before 1778 when Linnaeus died and probably before 1753 when Linnaeus published most of his binomial names. Hence, if one turns to H. E. Richter's *Codex botanicus Linnaeanus* (1835–40), one can soon find out, from entry 4829, what Linnaeus recorded about it and can learn more of its history. Publication of the Linnaean binomial goes back to Linnaeus's *Species Plantarum* (1753), a work which is the internationally accepted starting point for our modern scientific plant-names. Here Linnaeus adopted, however, the epithet *matronalis* from Dodoens, *Stirpium Historiae Pemptades:* 161 (1583) who called the plant *Viola matronalis,* this being his Latin rendering of the French names *Violetes des dames* and *Giroffees des dames;* these refer according to Gaspard Bauhin to the cultivation of the plant in gardens by ladics: 'Viola Matronalis quod a Matronis in hortis colatur dicta'. Gerard in 1597 called it 'Dames Violet'. A more important matter than the history of such a name as *Hesperis matronalis* is its precise application. Taking Linnaeus's publication as the first legitimate use of the name, we have to ascertain what plant he had primarily in mind when drafting his diagnosis; in other words, to typify the name. In almost all Linnaeus's concepts there is a definite primary visual element, either a specimen or an illustration from which he obtained such information as is summarized in his diagnosis, *Hesperis caule simplici erecto, foliis ovato-lanceolatis denticulatis* etc. This species was grown in Clifford's garden in Holland about 1736 and is well represented in the Clifford Herbarium now preserved in the Department of Botany, British Museum (Natural History), London; one of these Clifford specimens has been selected as the lectotype of *Hesperis matronalis* and provides a fixed reference point.

Nomenclatural types

The above exemplifies a further principle (No. II) of nomenclature: '*The application of names of taxonomic groups is determined by means of nomenclatural types*'. This seeks to ensure precision in the application of names by permanently associating them with 'types', which are occasionally illustrations as regards some Linnaean names but are specimens in herbaria as regards almost all others. Sometimes, of course, a type-specimen has been destroyed; unfortunately thousands of irreplaceable type-specimens went up in flames when the Botanisches Museum at Berlin-Dahlem was destroyed in an air-raid in March 1943. Sometimes, for new species described by some gardeners, it was left on the living plant! Thus type-specimens do not exist as vouchers for all botanical names. At times, when they do exist, their careful examination, in the light of present knowledge, may have unpleasant results: it may reveal, for example, that the original specimen differs from the plant now known under that name, and that the name has been misapplied. One such example is *Pyrus japonica,* later *Cydonia japonica* and *Chaenomeles japonica.* The name *Pyrus japonica* was published in 1784 by Thunberg for a dwarf shrub he had collected in a wild state on the Hakone mountains of Japan. Introduction of wild Japanese plants into European gardens was then very difficult and this Japanese quince did not reach them until about ninety years later; it was thereupon, i.e. in 1874, named *Pyrus maulei.* Long before then, however, a taller and more decorative Chinese species had been introduced into European gardens and identified, reasonably but incorrectly, as the obscure *Pyrus japonica.* Examination of Thunberg's specimen at Uppsala showed that his name *Pyrus japonica* belonged to the Japanese plant later named *Pyrus maulei* and the epithet *japonica*, whether the plant is placed in *Pyrus, Cydonia* or *Chaenomeles*, must be kept for this. Since two different plants must not bear the same name, another name had to be found for the Chinese plant, which is properly to be called *Chaenomeles speciosa*.

Priority of publication

During Linnaeus's lifetime his two-word names ousted the many-word diagnostic names which he himself continued to draft all his life and they become instead the standard names. Thereafter the binomial method became the only method of naming scientifically the species of animals and plants; it is obligatory under article 23 of the 1972 botanical code and article 5 of the 1964 zoological code. As, however, more and more botanists in different countries described and named plants new to them, it happened inevitably that the same plant on occasion received different names. Moreover, with a wide-spread species having different geographical races, that inhabiting one region would receive a different specific name from one inhabiting another region; thus members of the *Monadelphum* group in the genus *Allium* have received the names *A. atrosanguineum, A. monadelphum, A. fedtschenkoanum* and *A. chalcophengos* based on specimens from Dzungaria, Dauria, Turkistan, and western China which now seem to be conspecific. Which name should be adopted? Between the years 1804 and 1808 cultivated forms of the Chinese tree-peony received the names *Paeonia suffruticosa, P. arborea, P. moutan* and *P. papaveracea*? Which should be adopted? The trumpet-lily with numerous leaves and deep sulphur-yellow flowers which ranges from Upper Burma to western China was named *Lilium myriophyllum* by Franchet from Chinese herbarium material and *L. sulphureum* by Baker from Burmese plants introduced into cultivation. Which name should be adopted? Such problems continually come the way of anyone writing a monograph, a revision or even a check-list. They became particularly evident to the great Swiss botanist A. P. de Candolle (1778–1841) when engaged on a series of revisions of whole families. He decided that the only fair and practical method of deciding between such competing names was on a basis of priority, which already was a generally accepted but irregularly applied principle, as is evident from contemporary correspondence, because priority was a matter of fact, although not always easy to ascertain, and hence was not subject to personal and national prejudices and tastes. As de Candolle wrote in 1813, 'un nom ne doit point être change, parce qu'il est peu significatif; car on en trouve un troisième meilleur .... L'auteur même qui a le premier établi un nom n'a pas plus qu'un autre le droit de la changer pour simple cause d'improprieté. La priorité, au contraire, est un terme fixe, positif, qui n'admet rien d'arbitraire, ni de partial'. Thus the priority method prevented arbitrary name changes such as Salisbury made on a grand scale in his *Prodromus Stirpium* (1796); here he altered several hundred names to others he thought more apt. In the genus *Silene,* for example, he changed *S. anglica* to *S. arvensis; S. fruticosa* to *S. carnea; S. undulata* to *S. tristis; S. armeria* to *S. glauca; S. behen* to *S. ambigua; S. acaulis* to *S. caespitosa.* These new names are illegitimate under the *International Code,* for, if Salisbury were permitted to make such changes, why should not some-one else later change Salisbury's names for yet others seemingly more apt? The result would simply be nomenclatural instability, if not chaos. For such good reasons, principle III of the *International Code* reads: '*The nomenclature of a taxonomic group is based on priority of publication*'. Linked with this is principle VI: '*The Rules of nomenclature are retroactive unless expressly limited*'.

Thanks to these principles, thousands of names have been protected and continuity in their usage has been assured. Like most good rules, rigid insistence on priority has its disadvantages and these tend to be so exaggerated by gardeners as to cause them to overlook the much greater advantages. For example, the investigation of a group sometimes reveals the existence of an ignored name published earlier than the one in general use. According to principle III such an earlier name should be adopted. When, however, this affects the name of an economically important or well-known genus such as *Dendrobium, Calanthe, Nothofagus, Chimonanthus* or *Tectona,* the situation may be saved, and usually is, by getting the threatened name listed by an international botanical congress as a conserved generic name, a *nomen genericum conservandum.* An enormous number of name changes have thereby been prevented; the number of conserved generic names is now well over a thousand. The procedure for making such exceptions to the rule or priority involves the preparation of a detailed, well-documented and cogently argued case setting out the reasons for conservation of a given name and selecting a type-species: this proposal has then to be published and submitted to an international committee which will in due course recommend or reject its conservation. It costs many busy, expert people much time to examine critically such proposals and to report upon them. For a number of reasons, various proposals for the conservation of specific names have been rejected at successive international botanical congresses and there would seem no possibility whatever of future efforts being any more successful than those we have already made. Thus when such a name as *Paeonia albiflora* published in 1788 is found to be antedated by *P. lactiflora* published in 1776, all one can do now is to accept the earlier name with a good grace. Similarly, when the specific name of a well-known plant such as *Viburnum fragrans* Bunge (1833) is found to be duplicated by one published earlier, e.g. *V. fragrans* Loiseleur (1824) for a quite different plant and hence must be rejected as a later homonym, there is unfortunately no legitimate alternative to adopting another name for it, in this instance *V. farreri,* Changes of this kind tend to receive much adverse publicity. In fact.

however, the number of such changes due to priority is remarkably low in relation to the total number of plant names involved. Many more changes are due to differing views on the definition of genera and species. Uncertainty here may be due to inadequate knowledge of large and complex groups, which require monographic treatment, for example, the *Justicia* group in the *Acanthaceae* including *Beloperone, Adhatoda, Drejerella, Dianthera* and *Nicoteba*. Contrary to the opinion of many gardeners, very few botanists like changes in names because, apart from inconvenience, every change of name makes less accessible the information recorded under another name and so may lead to valuable information being overlooked. Thus, although it might appear easy, say, for a specialist in *Acanthaceae* to keep abreast of changes in classification of this large family, he would undoubtedly find it very difficult if not impossible to maintain a like acquaintance with, say, changes in the names of fungi, algae, lichens and mosses. Nomenclatural changes are particularly irksome for people engaged in ecological studies or even in editing papers for biological journals, but some are unavoidable.

Duplication of Botanical and Zoological names

There remains for brief consideration principle I of the botanical code which proclaims that '*Botanical nomenclature is independent of zoological nomenclature*'. Article 2 of the zoological code likewise emphasizes its own autonomy: 'Zoological nomenclature is independent of other systems of nomenclature in that the name of an animal taxon is not to be rejected merely because it is identical with the name of a taxon that does not belong to the animal kingdom'. This means that the same name can be legitimately used both for an animal and a plant in accordance with the relevant codes of nomenclature. The number of named species of animals, at least 1,120,000, and of plants, about 360,000, is so great that prohibition of name duplication would be disastrous. Efforts to avoid duplication would be extremely time-consuming, because not all of the immense literature of systematic biology is adequately and conveniently indexed; moreover the discovery of duplication would lead to very many changes in established nomenclature, bringing little if any good, but undoubtedly causing much inconvenience to botanists and zoologists alike. Indeed their activities and publications now overlap so little that they are mostly unaware of even the present extensive duplication. This rarely, if ever, causes confusion. Thus, although it is conceivable that the bird *Arenaria* (turnstone) might be found nesting on a clump of the plant *Arenaria* (sandwort) or that the butterfly *Pieris* be found sitting on the shrub *Pieris* or the butterfly *Fritillaria* on the bulbous *Fritillaria,* it is unlikely that the crustacean *Actaea* would have anything to do with the plant *Actaea,* the annelid *Dracunculus* with the plant *Dracunculus* or the orthopteran *Iris* with the plant *Iris.* For a botanist *Arenaria, Oenanthe* and *Prunella* are fairly well-known genera of British flowering-plants.

For the zoologist their names refer to nothing of the sort. In ornithology the generic name *Allenia* refers to a mockingbird, *Arenaria* to a turnstone, *Batis* to a flycatcher, *Cyanotis* a tyrant flycatcher, *Eremophila* a skylark, *Glaucidium* a pygmy owl, *Loddigesia* a hummingbird, *Macgregoria* a bird of paradise, *Newtonia* a fly-catcher, *Oenanthe* a wheatear, *Palmeria* a honeycreeper, *Passerina* a cardinal-grosbeak, *Prunella* a hedge-sparrow or accentor. For a botanist these are the names of plants. To an entomologist, *Agapetes, Clusia, Douglasia, Drymonia, Eucharis, Fumaria, Lasiopogon, Meriania, Microdon, Ricinus, Schrankia,* and *Stelis* are the names of genera of insects occurring in the British Isles, not of flowering-plants. Confusion through duplicated botanical and zoological names is likely to be troublesome only in the study of micro-organisms. Thus within the same handful of soil or farmyard manure there may live co-operatively together fungi and algae named according to the botanical code, bacteria named according to the bacteriological code and protozoa named according to the zoological code, all quite oblivious of nomenclatural complications. Hence the bacteriological code states as its principle 2 that 'The nomenclature of bacteria is independent of botanical nomenclature, except for algae and fungi, and of zoological nomenclature, except for protozoa'.

Utility of code of nomenclature

All codes of biological nomenclature, whether they concern botany, zoology, bacteriology, virology, horticulture, agriculture or forestry, have the same basic aim of promoting uniformity, precision and mutual intelligibility in the use of the names of organisms. They can only achieve this if their principles and consequent rules and recommendations are acceptable to the majority of people concerned with these organisms. Hence they have not been arbit-rarily imposed. They have been drafted and revised, circulated and debated, changed and edited in a remarkably democratic and international way by experts painfully conscious of their responsi-bility to their fellow workers and of the complexity of the subject, in order to make these codes both useful and acceptable; without them there would now be chaos, productive not simply of inconven-ience but of grave financial loss in many activities.

As botanists and gardeners using scientific names for our plants we are privileged to belong to a language community which from a small group of learned 16th-century apothecaries and physicians in southern, western and central Europe has grown in to a vast assemblage spread over the whole earth. The principles regulating the names of plants, which have been set out above, have their counterparts in the codes regulating the names of other organisms. The technical details of these codes and their different wording reflect the specialized outlooks of workers on different groups of organisms and also, it would seem, their different traditions. As regards the latter, the zoologists have thought it necessary to append to the *International Code of Zoological Nomenclature* a Code of Ethics. Article 2 of this insists that 'Intemperate language should not be used in the discussion of zoological nomenclature, which should be debated in a courteous and friendly manner'. For botanists and gardeners, who, as is well known, spend most of their time in calm contemplation of pretty flowers instead of ferociously dissecting fleas, sharks, and rats as do their zoological colleagues, such admonition is assumed to be quite unnecessary.

SOME SOURCES OF FURTHER INFORMATION

ASHWIN, M. B. (1958). Understanding plant names and their changes. *Tuatara* 7: 84–96.
BLUNT, W. & STEARN, W. T. (1971). *The compleat Naturalist: a Life of Linnaeus.* London.
CONGRESS; ELEVENTH INTERNATIONAL BOTANICAL CONGRESS, SEATTLE. (1972). *International Code of Botanical Nomenclature.* Edited by F.A. Stafleu & others. Utrecht.
CONGRESS; FIFTEENTH INTERNATIONAL CONGRESS OF ZOOLOGY, LONDON. (1961). *International Code of Zoological Nomenclature.* Edited by N.R. Stoll & others. London.
DILG, P. (1969). *Das Botanologicon des Euricius Cordus. Ein Beitrag zur botanischen Literatur des Humanismus. Inaugural-Dissertation, Philipps-Universität Marburg.* Marburg.
ELCOCK, W. D. (1960). *The Romance Languages.* London.
GLEASON, H.A. (1935). The reason behind scientific names. *J. New York Bot. Gard.* 36: 157–162.
———— (1947). The preservation of well known binomials *Phytologia* 2: 201–212.
GREEN, M. L. (later M. L. SPRAGUE). (1927). History of plant nomenclature. *Bull. Misc. Inf. Kew* 1927: 403–415.
KEYNES, J. M. (1906). *Studies and Exercises in Formal Logic.* 4th ed. London.
LINNAEUS, C. (1938). *The 'Critica botanica' of Linnaeus* [1737]. Transl. by Arthur Hort, revised by M. L. Green. London.
MCVAUGH, R. & OTHERS. (1968). *An annotated Glossary of botanical Nomenclature.* Utrecht.
RAIZADA, M. B. (1968). What is there in a name? Why change? *Indian Forester* 94: 37–46.
SAVORY, T. H. (1962). *Naming the living World.* London.
SMITH, A. W. & STEARN, W. T. (1972). *A Gardener's Dictionary of Plant Names.* London & New York.
SPRAGUE, M. L. (*née* GREEN) & OTHERS. (1944). A discussion on the differences in observance between zoological and botanical nomenclature. *Proc. Linn. Soc. London* 156: (1943–44): 125–146.
STAFLEU, F. A. (1956). Nomenclatural conservation in the Phanerogams. *Taxon* 5: 85–95.
———— (1971). *Linnaeus and the Linnaeans.* Utrecht.
STEARN, W. T. (1965). *Viburnum farreri,* a new name for *V. fragrans* Bunge. *Taxon* 15: 22–23.
———— (1973). *Botanical Latin,* 2nd ed. Newton Abbot.
———— (1967). Biological nomenclature. *Chamber's Encyclopaedia: New Rev. Ed.* 2: 324.
TURNER, W. (1963). *Libellus de Re herbaria,* 1538. *The Names of Herbes,* 1548. Facsimiles with introductory matter by J. Britten & others. London.
VOSSLER, K. (1932). *The Spirit of Language in Civilization.* Transl. by O. Oeser. London.

DISCUSSION

Mr J. CODRINGTON asked when Cotoneasters ceased to be ladies and became gentlemen, and why.

Dr STEARN replied that this was part of the business of trying to attain uniformity. The International Code stated that classical names taken into use retained their classical gender. The termination "aster" in Latin was always masculine and had a derogatory meaning—*Pinaster* was the kind of pine *(Pinus)* that did not produce the seeds you ate, poetaster was a rather inferior sort of poet; hence *Cotoneaster* was an inferior sort of *Cotonia* or Quince. The term *Cotoneaster* was therefore masculine for pre-Linnaean botanists and under the Code it should therefore be masculine. But there was a contradiction in that a very large number of trees like *Quercus, Pyrus* and so on were for some reason or other feminine for the Romans. Thus there was a conflict between a thing being a tree and a thing being an inferior sort. This, Dr Stearn thought, was why so many people at a certain period made *Cotoneaster* feminine, but in accordance with the Code it should be masculine and practically everybody was now making it masculine. There were a number of genera where the gender had fluctuated (for example *Panax*) but it made for uniformity to follow the Code. More people should read the little note in the Bulletin of the British Museum (Natural History) explaining this point [*Bull. Brit. Mus. (Nat. Hist.) Bot.* 1 : 125. 1954, –Ed.].

PROBLEMS OF HORTICULTURAL NOMENCLATURE

C. D. BRICKELL

Royal Horticultural Society's Garden, Wisley

The Codes of Nomenclature

Plants grown by man may (like Gaul) be divided roughly into three parts. Firstly there are the numerous species brought into our gardens from the wild, which are generally indistinguishable from their natural counterparts unless modified by many years of cultivation; secondly there are the selected variants and hybrids, often of complicated parentage which man produces and maintains for a wide variety of purposes; and thirdly there is a ragbag of unknown or unrecorded origin, some of which like *Peperomia caperata,* a well-known greenhouse plant not at present recorded in nature, have been described as botanical taxa from cultivated material. The naming of the first group is controlled by the *International Code of Botanical Nomenclature* and in gardens such plants should be given the names applied in the wild to the same taxa. The second and third groups are, in the main, subject to the *International Code of Nomenclature of Cultivated Plants* (1969), unless considered to be deserving of botanical rank, as in the case of *Peperomia caperata* (Yuncker, 1957) cited above. The "Cultivated Code" is a slim document compared to the Botanical Code previously mentioned, but to gardeners is, or should be, of equal importance, governing as it does the naming of many of our garden plants. It is itself subject to the rules laid down under the Botanical Code regarding the use of botanical names in Latin for both wild and cultivated plants, with the exception of graft hybrids. It is not, however, my intention to discourse on the Code in detail but to expand on some of the problems which those who grow plants face if they like to be accurate as to the names they use.

I would stress first of all that a primary purpose of both Codes is to stabilise the naming of plants. Simplicity of rules is very desirable in order to encourage their use, particularly in the case of the Horticultural Code, which gardeners are not obliged to use— and certainly will not if expert interpretation is required in order to apply the rules.

I would contend that, at present, there is insufficient linkage between the two Codes to allow a smooth, or relatively smooth, cross-over where the two disciplines of Botany and Horticulture meet. There is a tendency, on both sides, to sweep under the other's carpet, borderline problems which have, at present, no satisfactory pigeon-hole under either Code. Jeffrey (1968) concludes that "while formal botanical categories can successfully accommodate both wild and cultivated plants down to species level, they are inapplicable to the infraspecific levels of the systematics of cultivated plants".

I find myself disagreeing to some extent with his conclusion which appears to me rather sweeping and based too specifically on highly developed crop plants such as melons, bananas and wheats, without sufficient consideration of the many problems which occur with cultivated ornamentals, particularly trees and shrubs.

The use of the categories 'forma' and 'cultivar'

Many authors have considered that minor variants of a species should *not* be treated botanically. One has to face the fact, however, that a host of variegated or coloured-leaf variants, cut-leaved forms and numerous other variations within the species in respect of flower-colour, habit and so on already *have* been treated infraspecifically. In the past Rehder, Hylander and others have made use of what one may term the lower echelons of taxonomy, placing under the infraspecific categories *varietas* and *forma,* many of the minor departures from the normal form which occur both in the wild and in cultivation.

A current view of many modern taxonomists is that *subspecies* and perhaps to a lesser extent *varietas* are the only infraspecific categories which need be recognized taxonomically. This is very evident when one considers recent standards such as Davis's *Flora of Turkey* (1965–72) or many recent monographs. If the category *forma* is gradually being discarded by botanists it would seem not unreasonable for it formally to be used on some occasions as the link point for certain of the difficult cases which waver between the two Codes.

Jeffrey (1968) suggests further horticultural categories, *provar* and *convar* to accommodate the present limbo between species and cultivar but I doubt that the provision of additional groupings such as these would have the desired effect, and, I feel, would be more likely to confuse still further those who have to *use* the names and terms.

The category *forma* is or has been recognized and used by both botanists and horticulturists. Many ornamental plants named in the wild at this taxonomic level are cultivated. As an example, *Carpinus betulus* L. forma *quercifolia* (Desf.) Schneider has been described to include variants of the common hornbeam with oak-like leaves. Such trees are not known to occur as distinct populations but are said to be found sporadically in wild stands as individuals. If this variation is considered worthy of recognition (which some taxonomists would of course dispute) the epithet provided by Desfontaines and accepted by Schneider is valid to provide a reasonably satisfactory identity-tag for them. Assuming that one or more of these individuals is brought into cultivation it seems pointless, and indeed misleading, to treat them not as forma *quercifolia*, a slightly variable but reasonably circumscribed entity, but as 'Quercifolia', as would be recommended under schemes insisting on the separation of wild and cultivated plants at infraspecific level. Written in this way it will generally be accepted (incorrectly) as a clone of uniform leaf-shape. This results from the terms cultivar and clone, although clearly differentiated in the Horticultural Code, being confused, misunderstood and frequently equated.

The difficulty arises from the fact that *all* cultivar categories (four are cited in the 1969 Code) are written in a similar manner using a capital initial letter and enclosing the epithet in single quotes. There is a failure therefore to distinguish between these categories and it is not possible in print to tell whether they are increased by vegetative or asexual means (clonal) or by seed, and are then possibly variable (unless apomictic) within the limits laid down. In the case of ornamental plants, annuals generally excepted, this lack of differentiation is important commercially. This I feel is a problem the Horticultural Commission must grapple with fairly soon before the category 'cultivar' becomes devalued by continual misinterpretation of its functions by horticulturists who do not understand its present limitations and by botanists who use it (incorrectly in my view) as a convenient receptacle for variants which are bothersome to place using conventional taxonomic practice.

One of the arguments against according botanical rank to such variations is that they are often transient and unless maintained artificially by man will die out leaving only an unwanted epithet to clutter yet further the literature. This argument is certainly valid in the cases of freaks such as witches brooms on conifers and although in the past these variants have been given latinized names—by both botanists and horticulturists—they should undoubtedly be given clonal names, and not botanical rank.

But there are many instances where the status is not clear cut. The case of the oak-leaved hornbeams has been cited previously. An example (at slightly elevated taxonomic level) is that of the purple or copper beech *Fagus sylvatica* L. var. *atropunicea* Weston, which is recorded wild from France, Switzerland and Germany. It varies in depth of foliage colouring from green through the palest copper to the deepest black-purple if raised from seed and is a very popular garden plant. If treated in gardens as a cultivar, 'Atropunicea', this name would still be required to cover all seed-raised plants with this range of purple or copper foliage colour. It might at a pinch be considered a cultivar as defined in the Horticultural Code, Article 11, category C. "A cultivar consisting of cross-fertilised individuals which may show genetical differences but having one or more characters by which it can be differentiated from other cultivars". But it does not fit properly into this category as here defined. Bearing in mind the clonal interpretation most gardeners now place on cultivar names, particularly for woody plants, the confusion which can and does occur is obvious. A gardener might order from a catalogue *Fagus sylvatica* 'Atropunicea', described as "the copper beech, beautiful copper-purple foliage" and expect to receive a plant with deeply coloured purple-brown leaves. He may receive such a plant; but he could also be sent, quite legitimately, any variation of this colouring pale or dark, should the cultivar name be applied in this way. It would seem reasonable, therefore, in cases of this type to make use of a well-established category which is generally understood rather than to create further horticultural categories which are unlikely to become established and will certainly be misunderstood.

One could also have a case of naturally occurring copper beeches, being brought into cultivation and in order not to contravene the first section of Article 28 of the Botanical Code which states that "Plants brought from the wild into cultivation retain the names that are applied to the same taxa growing in nature" they would be designated as var. *atropunicea*. The situation becomes slightly ludicrous if then one also has in the same garden or nursery cultivated copper beeches labelled 'Atropunicea'! The whole exercise becomes confused, cutting across the basic concept of stabilization of names. Here we have a situation where both the botanical epithet and the cultivar name could refer to two groups of plants which can only be distinguished if one happens to know the origin of each group. It seems immaterial from the practical viewpoint to attempt to maintain such artificial and unnecessary distinctions, particularly when a perfectly adequate name exists to be used. And yet this is the policy "separatists" among the two disciplines advocate!

A further example is that of *Juniperus communis* L. var. *suecica* (Mill.) Ait. This name has been applied to columnar plants occurring wild in Scandinavia and E. Prussia (Dallimore, Jackson & Harrison, 1966) and also to plants cultivated as the Swedish Juniper in Britain, now often designated 'Suecica'! These cultivated plants do not appear to be derived from a single individual and similar plants occur not infrequently in Sweden and elsewhere (Hylander, oral comm.) This case is parallel to that of the copper beech and the use of the botanical trinomial would seem logical. This does not preclude an individual within var. *suecica* (or any other infraspecific category) being selected for a particular character, or characters, propagated and given a cultivar (clonal) name, but the "umbrella" of var. *suecica* is still available for the remainder, whether wild or cultivated.

Jeffrey (1968) states that "botanical infraspecific classification depends upon the formal recognition of natural populations of various degrees of distinctness and genetic isolation", and considers that as cultivated plants do not exhibit a natural population structure they should be named by a system independent of infraspecific botanical nomenclature.

It may well be that this concept is a desirable one from a purely detached consideration of the needs of botanical taxonomy but is it a realistic and practical assessment of the problems? Accepting infraspecific categories to recognize natural populations alone as this seems to imply will exclude from any recognition naturally occurring minor variants of flower colour, foliage variegation and similar departures from what is accepted as typical for an individual species, if they do not fall within the limits quoted above. Some may form distinct populations in a particular area; others will occur sporadically as individuals within the range of the species, as in *Colchicum speciosum* where white-flowered forms may be found dotted around in natural wild stands of the species. Why should not this sporadically occurring variant be considered equally as worthy of recognition, as for instance variants distinguished (in very recent floras) by relatively minor characters not necessarily associated with geographical distribution or genetic isolation. Often variants like white-flowered forms of *Colchicum speciosum* are brought into cultivation from wild stands and no name is available to use for them. If only a single clone is involved no difficulty need arise; a cultivar (clonal) name is coined and a satisfactory "handle" has been provided. But where there is variation and that variation can be reasonably circumscribed as in the case of the purple beech previously cited, I would suggest that a link-point could be established at *forma* level without seriously affecting the botanical classification of infraspecific groups.

Why for instance should wild stands of *Tulipa clusiana* DC. var. *chrysantha* (A. D. Hall) Sealy, a mere colour-variant of the type species be categorized as 'Chrysantha' when brought into gardens—a logical alteration if one accepts that infraspecific variations of wild and cultivated plants should be classified under different systems. Why also should *Ursinia chrysanthemoides* (Less.) Harv. var. *geyeri* (L. Bolus & H. Hall) Prassler, a colour-variant confined apparently to the Ceres district of Cape Province be treated as a cultivar, 'Geyeri', when brought into cultivation. Horticulturists are by now well used to dealing with trinomials (and polynomials!) and it very confusing to gardeners when the most recent monograph of a genus (Prassler, 1967) accords botanical rank to this wild variant and yet a few years later, and dealing with this individual taxon in isolation (Jeffrey, 1970), it is deemed to be a cultivar.

Only cases comparable to the examples I have cited would require the type of treatment I have advocated. Each individual case would need to be considered and documented carefully. There is of course nothing (except possibly publication space!) to prevent any individual providing valid botanical epithets in the categories *forma* and *varietas* and Rehder (1949) did just that for many woody plants. Far too many infraspecific epithets have already been published, however, with brief, often poor descriptions, and for any botanist or horticulturist to add to the number without very careful consideration would be irresponsible. Lack of any coherent policy in infraspecific classification has led to what Burtt (1970) vividly describes as a "muck-heap of two centuries of unindexed and inadequately described epithets".

I am certainly *not* advocating that all those garden plants which in the past have received pseudo-botanical Latinized names should be recognized botanically. The majority can quite readily be shown to be clones and should be treated accordingly.

I am well aware, of course, that these arguments may not be entirely applicable when dealing with highly developed crop plants but the views are put forward in the hope that wider discussion and cooperation between botanists and horticulturists may produce a scheme, or schemes, acceptable to both. Whether we like it or not botany and horticulture are inextricably intermingled in literature and gardens (even botanic gardens!), and to carry out an isolationist policy, as is sometimes advocated, can lead to even worse chaos than at the present time.

Name changes

In the thorny question of name changes as they affect horticulturists, it often seems to gardeners that the Latin names of plants one knows well have been or are about to be changed for no good reason. There are, of course, excellent reasons in most cases. Gardeners frequently bring back plants from visits abroad and unless they obtain specialist identification may well grow and distribute widely a particular species under a totally erroneous name. The case of *Euryops acraeus* M. D. Henderson, a small South African shrub, brought into cultivation about 1946, illustrates well the necessity for correcting a mis-applied name. Early in its garden career *E. acraeus* was mis-identified as *E. evansii* Schlechter, a closely related but distinct species and so was (and undoubtedly still is) grown under this name. It was later realized the the plant in cultivation was not *E. evansii* but *E. acraeus,* first described as a new species in 1961 (Brickell, 1970).

Another South African plant misidentified in our gardens for many years is *Osteospermum* (*Dimorphotheca*) *jucundum,* generally grown as *Dimorphotheca* or *Osteospermum barberae,* a similar species (Marais, 1971).

Taxonomic changes due to the re-assessment of a plant's status in the light of additional knowledge also occur. The well-known Shrimp plant cultivated as *Beloperone guttata* has had a chequered career in this respect. It has been included in *Drejerella* as *D. guttata* and seems at present to rest as *Justicia brandegeana.*

Alterations of this type are made after very careful study of the relationships of the plant concerned and, with a reasoned explanation, are gradually absorbed into the horticultural scene and accepted. After all, gardeners will argue vehemently over correct methods of cultivating particular plants and as new knowledge and techniques come along, quickly bring them into use. So it is not unreasonable for them to allow botanists the same privilege with taxonomic matters.

Much more frustrating to gardeners, and also to many botanists are name-changes for purely nomenclatural reasons, based on the principle of priority. As an example, few garden plants are better known than *Viburnum fragrans,* yet this is an illegitimate name for the superb winter-flowering shrub known to thousands of gardeners. Unfortunately, the name *V. fragrans,* given to this species by the botanist Alexander von Bunge in 1833, is ante-dated by Loiseleur's use in 1824 of *V. fragrans* for a different plant, and under the rules of botanical nomenclature the earlier published name has priority. Bunge's use of the name must be rejected and in 1966 Dr W. T. Stearn reluctantly published the name *V. farreri* for this species. Stearn has been aware of the duplication of the name *V. fragrans* for over 20 years but had delayed publishing a new name for this well-known plant in the hope that proposals for some form of conservation of specific epithets would be adopted by International Botanical Congresses. This, alas, has not occurred. The curious rigidity of the rules is, in some respects, hard for gardeners (and many botanists) to understand.

Under the Botanical Code, conservation of a fairly large number of generic names is already accepted, and after reasoned argument others are added to the list at each Botanical Congress It would not appear unreasonable for a similar procedure to be adopted for specific epithets, again after arguments had been put forward for consideration by a specialist committee and accepted or rejected as befitted the case.

Few people will see reason in altering the botanical name of the tiger lily, universally known as *Lilium tigrinum,* to *L. lancifolium,* but under the rules as at present formulated this regrettably should be done.

Generally, of course, the effect of applying rules accepted internationally is extremely beneficial in stabilising plant names, always bearing in mind that natural entities do not necessarily lend themselves to mathematical treatments!

Support for the conservation of specific epithets is not entirely horticultural. Brummitt (1972) has shown that the correct name for the plant well-known as *Galax aphylla* L. is *G. urceolata* (Poir.) Brummitt, a change brought about for purely nomenclatural reasons. In his final comments he suggests that botanists might well consider again the desirability of conserving specific names, a suggestion with which I find myself in complete agreement One cannot tell how many further names are still to be upset when their nomenclature is studied in detail. Apparently usage and the importance in industry, horticulture, agriculture and forestry of maintaining reasonable stability of specific names are factors of small account when such matters are considered. Yet there seems little danger in swamping the market with conserved specific names if a watchdog committee of the calibre of those who deal with conservation of generic names is appointed.

Another sad case affecting gardeners is that of the plant long known as *Hosta albomarginata,* which if the Botanical Code of 1972 (Article 75) is adhered to strictly, should now be called *Hosta sieboldii,* in spite of the existence of the valid application of the name *Hosta sieboldiana* to another species. Hylander (1954) pointed out many years ago that *siedoldii* (based on Paxton's *Hemerocallis sieboldii*) was the prior epithet for *H. albomarginata,* but very sensibly considered it would only complicate further an already complicated nomenclatural situation. The combination *H. sieboldii* (Paxton) Ingram has now been made for this plant, the author expressing the somewhat curious view at one point in his argument that confusion is unlikely to arise as the two *plants* are so very different (Ingram, 1967)! That the two species are distinct no-one would dispute. Distinctness of plants is no guarantee that confusion will not occur when the names are so similar—as anyone involved with the practical application of names to plants is only too well aware.

It seems curious that the Botanical Code permits the use of two such similar specific epithets as *sieboldii* and *sieboldianus* within the same genus. In fact, in the Code an example of epithets *not* likely to be confused which is given in Article 75 is the parallel case of *Lysimachia hemsleyana* and *L. hemsleyi*! However, in another part of the Code, under Recommendation 23A we read that "It will be well, in the future, to avoid the use of the genitive [in this case *sieboldii*] and the adjectival form [*sieboldianus*] of the same word to designate two different species of the same genus". Let us hope that the next edition of the Code will make this a firm rule and that the rule will be retroactively applied.

Problems of this type abound. How, for instance, is one to develop a system which distinguishes, within the concept of a broad species, reference points to horticulturally important "species" which may now be included within it. An example is provided by *Caryopteris incana* (Houtt.) Miq. which is now held to include *C. tangutica* Maxim. As known from Farrer's collection in Kansu (*F.* 305) this latter plant is hardier and flowers earlier than *C. incana* as grown in gardens. A cultivar name is not really appropriate. Burtt (1970) suggests bracketing the "sunk" specific epithet to follow the species name, which offers a possible solution. The problem becomes much more acute in *Rhododendron* and the wrath of a Rhododendron enthusiast about to "lose" a species has to be felt to be believed! It would seem very desirable to provide and recognize a system to deal with this point, for botanists as well as horticulturists sometimes require to refer to taxa which have been "sunk".

Multiplicity of cultivar names

The almost uncontrolled rash of cultivar names, some correctly applied, others unwarrantably provided is difficult to rationalize, although the establishment of International Registration Authorities helps considerably with certain genera. Some horticulturists tend to apply cultivar names indiscriminately without due attention to description, comparison, documentation and indeed the necessity for a name at all. If some curb is not placed on the process the chaos will be even worse than in the past. Two examples will suffice to show how confusion can easily occur. A *Sedum* species which I was given under the cultivar name 'Adolph Hamet' proved on investigation to be *Sedum adolphi*, a species described by the botanist Hamet! The author's name had in some curious fashion been incorporated into a pseudo-cultivar name. One can more and more find similar cases in the literature, particularly of quite legitimate infraspecific botanical epithets being given an initial capital letter and plonked into single quotes—and another unwanted cultivar name is born. Even legitimate cultivar names do not escape. A nurseryman friend provided me with a prime example for a Juniper cultivar known as 'Bonin Island'. When looking at his plants of this cultivar one day he found them labelled *Juniperus* 'Born in Ireland' and on enquiry discovered that the original label had been indistinct and that this was the interpretation of the lettering remaining on the faded label! Whilst such errors may be amusing they are symptomatic of a growing carelessness difficult to halt. Much thought I personally dislike official control, it may well be that the production of annual or biennial lists of names to be used obligatorily in catalogues will be necessary. In particular, I would suggest that no cultivar name should be given without the sanction of the appropriate Registration Authority, or some similar body if no Registration Authority for a particular genus be available. At present there is no legal pressure available to bring to bear on those who will not register cultivar names with the I.R.A. concerned. If registration of a cultivar name is legally enforcible before the plant can be sold and if registration involves time, trouble and a small monetary outlay, only really distinct cultivars are likely then to be introduced—and the cultivar names should then be correctly formed and applied. Many nurserymen and breeders do, of course register their plants correctly now, but there is very considerable room for improvement amongst others.

The need for co-operation

These then are outlines of a few of the problems involved. I feel strongly that both botanists (not merely botanical nomen-claturists) and horticulturists must re-examine the Codes and ask themselves what they really require from them. Control is essential. The present Codes provide rules to follow and in most instances they appear to be sensible and workable. But do they always provide the stability of naming we wish to find? Or are we tending to treat natural entities too mathematically and in doing so defeating partially our purpose of producing a precise but simple system? Even though we strive to produce natural systems of classification, our categories, being man-made, are artificial and will always be so.

Nature (fortunately) does not provide plants in neat compartments which do not overlap. Our systems must therefore be sufficiently flexible to allow for the overlap—but are they?

It has been, and still is, considered by some that there is no need for the two Codes to be on more than polite speaking terms with one another. My personal opinion is that they should be much more strongly related and both will benefit. With genera which are grown in gardens, co-operation between a botanist producing a monograph on a particular genus and the gardeners who grow the plants will often prove beneficial to both. Indeed for many petaloid monocotyledons a satisfactory and thorough taxonomic study cannot be undertaken unless the plants are grown for the botanist to see *in vivo*.

The gardener's expertize in cultivation, and often his or her observation of the plant's behaviour, can be important and could well be taken more into account by the monographer than in the past. Similarly gardeners should be encouraged to be more flexible in their attitude to botanists who to them appear to exist merely to change names! Gradually the various problems must be overcome—but they will take considerably longer to solve if the two disciplines do not try to co-operate more fully than in the past.

REFERENCES

BEAN, W. J. (1970). *Trees and Shrubs Hardy in the British Isles*. 8th Revised Edition, 1: London.
BRICKELL, C. D. (1970). *Euryops acraeus* or *Euryops evansii*? *J. Roy. Hort. Soc.* **95**: 94–95.
BRUMMITT, R. K. (1972). Nomenclatural and historical considerations concerning the genus *Galax*. *Taxon* **21**: 303–317.
BURTT, B. L. (1970). Infraspecific categories in flowering plants. *Biol. J. Linn. Soc.* **2**: 233–238.
DALLIMORE W., JACKSON A. B. & HARRISON S. G. (1966). *A Handbook of Coniferae and Ginkgoaceae*. 4th Revised Edition. London.
HYLANDER, N. (1954). The genus *Hosta* in Swedish gardens. *Acta Hort. Berg.* **16**: 331–420.
INGRAM, J. (1967). Notes on the cultivated Liliaceae 5. Hosta Sieboldii and Hosta Sieboldiana. *Baileya* **15**: 27–32.
JEFFREY, C. (1968). Systematic categories for cultivated plants. *Taxon* **17**: 109–114.
——— (1970). *Ursinia chrysanthemoides* cv. 'Geyeri'. *Bot. Mag.* **177**: t. n.s. 564.
MARAIS, W. (1971). The identity of *Dimorphotheca barberae* of gardens. *J. Roy. Hort. Soc.* **96**: 36–7.
PRASSLER, M. (1967). Revision der Gattung *Ursinia*. *Mitt. Bot. Staatss. Munchen* **6**: 363–478.
REHDER, A. (1949). *Bibliography of Cultivated Tree sand Shrubs*. Jamaica Plain.
STEARN, W. T. (1966). *Viburnum farreri*, a new name for *V. fragrans* Bunge. *Taxon* **15**: 22–23.
YUNCKER, T. G. (1957). New species in *Peperomia*. *Kew Bull.* **12**: 421.

DISCUSSION

Mr A. J. HUXLEY asked for comment upon the practice of the Royal Horticultural Society in giving clonal names to wild species put up for awards, notably in orchids.

Mr BRICKELL replied that he believed there were far too many cultivar names given at the least drop of a hat and he personally felt that quite a lot of the names which were provided, even to some of the plants which were given awards, were wrongly called cultivars. He saw no reason why a species, as such, should not be given an award when it was reasonably uniform. There had been a case where a clone of a *Dionysia* species had been given an Award of Merit. He personally thought that was incorrect and that it should not have been given a clonal name, and said so, but it still received one. He felt that unless a species varied quite considerably there was no justification for giving a clonal name. However, in the case of Rhododendrons and orchids, where individual species did vary an enormous amount, if one had a plant which was given an award, unless it was provided with a clonal name, a term he prefered to use instead of cultivar, that species could then be sold as having, for example, an Award of Merit. It might well happen that there were inferior forms of that same species in some nurseries and so one could quite legally have a nursery offering, say *Rhododendron arboreum* A.M., and they could be selling either a poor form or the good one. If the plant had been given a clonal name, only that individual and its vegetatively propagated offspring would then be entitled to that name, and if a nursery did not provide that particular clone under that name then one could presumably sue them. That was the reason for the practice, and with Rhododendrons and plants of that type one did need to distinguish between different variants for commercial purposes in order that the public who bought them were not misled.

Mr HUXLEY added that he thought there was still some confusion left, for the person who received a particular plant had no impression of the species as a whole. One could not tell whether the species itself was worth anything or just that example of it.

Mr BRICKELL said he thought this should be stated in the description, but it was impossible in the award system as it stood to give this information. However, in the description one could say "*Rhododendron arboreum* is a very variable species, there are white, pink and red forms of it. This particular clone is distinguished by...".

Mr HUXLEY added further that he thought the case of the *Dionysia* was in fact based on this degree of variation, the kind one saw in wild plants.

Mr BRICKELL replied that once one started, in a genus like *Dionysia,* to name clones, in theory one had to raise all the plants from cuttings. Also one ran into the danger, which he knew did occur in practice, that people named clones and then propagated them from seed and sold the offspring under those names. As an example one could take *Helleborus* 'Potter's Wheel'. One would not, of course, bother to raise most Hellebores by division; it took too long and commercially it was not an acceptable method. But one could still find *Helleborus* 'Potter's Wheel' sold under a clonal name. Yet this really was, or should be, one individual plant and its vegetative offspring. This was one of the confusions to be found in not distinguishing below the rank of cultivar.

Mrs M. GILLIAM commented that in her day-to-day work she needed to ensure that students were using the correct and up-to-date names. Could she be told of an easy reference for cultivated plants, other than woody plants (which seemed well served, or relatively so). How could they make quite sure they were using the most up-to-date name when, living far from London, they were generally out of touch with the Libraries and personalities who worked there and whom they could consult.

Mr BRICKELL replied that as far as he knew there was no reference book which would help.

Dr STEARN said that there was no absolutely good, up-to-date work. Zander's *Handwörterbuch der Pflanzennamen* (ed. 9, by Zander, Encke &

Buchheim, 1946), a little German work, was very good in general, but we had a lot of plants in cultivation in Britain which were not covered by it. Also there were a lot of cases where one just did not know what the right name was. The Shrimp Plant was a case in point. If people asked him what the correct name for it was, having studied the matter for a good many years he would say he did not know. He would say Shrimp Plant and *Beloperone guttata,* although he was quite certain that *Beloperone guttata* was not its right name. But here, in order to get the right name, one had first to study the pollen, as well as the floral characters, of some 300 species scattered over the world. A monograph of *Justicia* was really a five to ten year job and a working taxonomist just had not the time to do this in order to ascertain the correct name for a single garden plant.

THE DEVELOPMENT OF GARDEN PLANTS FROM WILD SPECIES

R. D. MEIKLE

Royal Botanic Gardens, Kew

The diversity of Nature

It will come as no surprise to field botanists, and gardeners, to learn that plants vary in a marvellous diversity of ways. Such variation provides the essential background to theories of evolution, to natural selection and the origin of species, theories which, in the penetrating glare of hindsight, seem hardly more than glimpses of the obvious.

Plants sport, interbreed, or become diseased; they throw up giants or dwarfs; they multiply various organs, or alter their shapes and structures, or change their colours. Sometimes the changes are so striking (as in the case of the peloric *Linaria vulgaris*) that botanists can be fooled into seeing them as new genera, or even as new families. But often the variations are so trifling that only after the most careful scrutiny does one appreciate that there has been any departure from the norm. So great indeed is the potential mutability of each individual plant that what astounds is not the effrontery of taxonomists in supposing that they can bring order out of vegetable chaos, but rather the extent of their success in doing so. For although no two individual plants, even of the same species, are ever exactly alike, yet, at the same time, one can travel the world over, and with astonishingly few exceptions, meet and recognize the same families, the same genera, and sometimes over very great areas, the same species. I am not astonished by the concept of evolution, but I am astonished by the fact that natural selection has not left a murkier trail of intermediate, connecting, divergent, ill-defined and misbegotten species, genera and families. It is the relative neatness, orderliness and discreteness of the vegetable world that astounds me. So, in addition to variation, one must admit the existence of a mechanism for stabilization, a mechanism for maintaining distinctions which is efficient enough to prevent constant change and a lapse into chaos, but at the same time sufficiently flexible to allow each individual a measure of liberty to alter its appearance or mode of life to a very small degree, and to allow the occasional plant to make a big and dramatic departure. It is in part this inherent potential for mutability which has permitted the development of garden flowers from wild species, but no less the potential for stability which has allowed us to maintain and consolidate such developments.

Nature may abhor a vacuum, but—of this I am convinced—in the world of plants, Nature even more abhors perfect mechanisms and absolute principles. No one plant mechanism seems to work perfectly; no one principle is absolute; no one departure from the usual is impossible. The progress of research is not the revelation of principles but rather the multiplication of exceptions and irregularities.

The first gardeners and their plants

Where does this get us? In the beginning Man, expelled from Eden, took to tilling the soil. And Nature, though rather frightful, was not wholly unco-operative. A diet of wild fruits, roots and leaves may not have been very exciting, but after the very stupid (who are generally the very courageous) had been eliminated by Hemlock, Deadly Nightshade and Death-cap, the not-so-stupid must have begun to realize that amongst single sorts of the innocuous kinds (I am avoiding the philosophical term species) some fruits were sweeter than others, some roots more succulent, even where the plants producing these fruits and roots were otherwise virtually indistinguishable. The most inexperienced Blackberry-picker soon learns this, and in doing so rediscovers the birth of gardening; for the mechanism of stabilization tends to insure that the fruit would remain sweeter and the root more succulent so long as it was gathered from the same individual, or from descendants or offsets from the same individual. From this observation, it was a logical and sensible step to safeguard the valued plant by removing it to the proximity of a dwelling, where it could be watched and propagated, and conveniently harvested in due season. History does not record the name of the first gardener, but I am prepared to controvert Genesis (and Tom Paine) and suggest that it was an Eve rather than an Adam. And in this, if my recollection serves me right, I have reason to believe that no less an authority on antique cultivation than Dr William Stearn is on my side. Is it not significant that the major deities most closely associated with the cultivation of plants— Cybele, Rhea, Demeter and Persephone—are, almost without exception, ancient and formidable females? And is it not significant that the Golden Age—that permanent yesterday—was succeeded by the Silver, when, according to some sources, a world of weak, submissive men was ruled by a powerful and omniscient race of women? If myths mean anything, then humanity owes a lot to these enterprising ladies, for without knowing the difference between a stamen and a stigma, they saved mankind from an eternity of chasing and being chased—at least by others than their own kind.

By the time history replaces myth, a surprising number of useful plants were not only in cultivation, but already highly developed, sometimes with quite an extensive range of cultivars. Theophrastus, the father of botany, who is said to have lived

from 370 to about 225 BC., was familiar with cereal crops and
practically every common vegetable we grow today except Sweet
Corn and Potatoes. He mentions Cabbages, Radishes, Turnips,
Beet, Lettuce, Leeks, Onions, Shallots, Cucumbers, Gourds, Peas,
Beans, Celery, Mustard and Cress and, amongst fruits, Almonds,
Cherries, Plums, Pomegranates, Apples, Pears, Quinces, Chest-
nuts, Dates, Hazels, Figs and, of course, Vines.

The list of potherbs given by Theophrastus would do credit
to the pages of Thompson & Morgan, but when it comes to ornamental flowers, we slip back through the centuries to the very
beginnings of decorative gardening. His flowers are not always
easy to identify with certainty, but we can be confident that he knew
of several kinds of Roses, of Lilies, Violets, Anemones, Narcissi,
Lavender, and possibly of Stocks, Wallflowers, some sort of Clove
Pink, and certainly of a sweet-scented *Artemisia,* perhaps Southern-
wood. All his flowers, except the Anemone, were valued primarily
for their scent, and for making garlands and nosegays, and all
had been derived from wild plants common in Greece and the eastern
Mediterranean. Not much of a catalogue, you may say, but it was
to serve Europe substantially unchanged for the next eighteen hundred years. Pliny the Elder, born some 300 years after Theophrastus,
adds Periwinkle to the list of garden flowers, but very little else, nor
is it always very clear from his diffuse and often rather obscure
Natural History whether he is discussing flowers that were cultivated,
or wild plants used in making garlands, perfumes and the like.

Now listen to what Sir Frank Crisp has to say about floriculture
in the Middle Ages: "wild flowers", he says "were the predominant
feature of Mediaeval gardens, and what in post-mediaeval times were
called 'Nosegaie flowers' were very limited in number—the Rose,
Lily, Violet, Clove-pink and Periwinkle". Almost a repetition of
Theophrastus and Pliny, with Roses, and possibly Clove-pinks,
the only plants represented by garden developments significantly
different from their ancestors. I suppose it might be argued that
the harsh conditions of mediaeval life and the general atmosphere
of insecurity were scarcely conducive to flower-gardening. That
may be so, but I also feel that the classical and mediaeval attitude
to plants in general, and to ornamental flowers in particular,
illustrates two distinct stages, or possibly phases, in human thought,
one of which may be called "natural and primitive", the other
"unnatural and advanced". To the majority who lived in the ancient
world of Greece and Rome, and to many who still live in undeveloped, or thoroughly rural parts of the present-day world, the
word "plant" means primarily a cultivated plant with a market
value—a cabbage or something similar, which can be sold for money,
or else a herb with widely accredited and highly valued medicinal
virtues. There are, in addition, two "flowers", the Rose and the Lily,
which by hallowed tradition symbolize the absolute in beauty and
purity. All the rest are "weeds", not because they invade the
garden or spoil the crops, but because they are unpleasantly reminis-

cent of Nature, and Nature is, at best, something which one would
rather not be reminded of. Occasionally a plant smells so sweetly
that it is elevated from the level of "weed" to the select company
of "flowers", and this advancement in rank is not infrequently
reflected in its popular name. Thus *Primula vulgaris* ceases to be
vulgar and becomes a Rose, while *Convallaria majalis* is designated
a Lily, though in fact neither bears the least resemblance to its name-
sake. Beauty alone is no guarantee of social success, unless that
beauty is so astounding ,or so alien to its environment, as to demand
recognition. *Narcissus pseudo-narcissus* is still called a Lent Lily
in some parts of Britain, while in rural areas of Cyprus, the Oleander
passes as a "Rose Bay" and *Anemone coronaria* and *Gladiolus
segetum* stand out as "Lilies" amongst a mass of equally colourful,
but otherwise insufficiently distinguished "weeds".

The "unnatural and advanced" mind is, of course, the converse
of all this, and is characteristic either of the privileged few, who,
thanks to wealth or position, have no need to struggle with Nature
and the soil, or of the modern urban masses, who, long divorced
from the wearisome round of rural toil, get their vegetables (washed)
in polythene bags at the supermarket and their milk in bottles on
the doorstep. To such as these, Nature normally presents a
smiling face, and things natural seem nicer and more wholesome
than the artificialities of organized city life. This benevolent
image of Nature is, I need hardly say, assiduously fostered by the
proprietors of countryside magazines, and is profitable, saleable,
acceptable—indeed only credible—in societies where wild-life, in
all its forms, is securely under the heel of modern technology.

I have mentioned Pliny the Elder—his nephew, Pliny the
Younger, a rich and powerful man, and a friend of the Emperor
Trajan, describes in some detail the gardens of his villa in Tuscany.
The villa was set in rich agricultural country, with vineyards and
cultivation all around; there were Bay trees, Planes, Cypresses,
fruit trees, Ivy, and spacious lawns in the garden, and plenty of
room for flowers, with no doubt a sufficiency of gardeners to tend
them. But Pliny mentions only two flowers, namely Roses and
Acanthus, for the rest (in his own words) it was "box shrubs clipped
into innumerable shapes, some being letters, which spell the gardener's name or his master's, small obelisks of box alternate with
fruit trees, and then suddenly in the midst of this ornamental scene
is what looks like a piece of rural country planted there. The
open space in the middle is set off by low plane trees planted on each
side; further off are acanthuses with their flexible glossy leaves, then
more box figures and names". The scene is central Italy in the
2nd century AD., but it might equally well be the garden of some
Tudor prince or magnate, nor do I think that the similarity is
attributable solely to the revival of Classical learning or the acceptance of Roman standards in matters of taste. Pliny, like his
Tudor counterparts, had every reason to enjoy the countryside:

it offered an escape from court intrigue: it provided the excitement of the chase; it furnished an income from rents and, best of all, it called for no more than a casual supervisory involvement in the arduous round of argicultural toil. Nature, under control, had such a pleasant face that little pieces of rural countryside fitted happily into the garden scene. The clipped obelisks and yards of topiary alone reveal a lingering element of primitive insecurity; they are the antithesis of naturalness, and represent the subconscious assertion of man's supremacy over the wild. The gentry of the 16th century still felt the need for such symbols of domination, but with growing confidence in the established order of things, the topiary diminishes, and the great man's garden assumes increasingly the aspect of natural informality. Flowers, in the popular sense, and flower-gardening were never in harmony with this concept of a paradise, nor do I believe that the magnates were normally enthusiastic participants in the florist's department of horticulture. Such unnatural productions, though on occasion useful as decorations for the house, were usually relegated, along with their associates—herbs and vegetables—to a walled enclosure, out of sight, and under the sole and exclusive supervision of the gardener and his assistants.

Fortunately for floriculture, the magnates did not alone have gardens; there were plenty of small enclosures where contrived rusticity and landscape gardening on the grand scale were neither feasible nor particularly admired. For the owners of such modest estates, the contemplation of sweeping vistas was out of the question, but satisfaction of a different sort could be had from the closer inspection of a flower bed where, at small cost, one could have riches in colours, scents and textures, outbidding even the magnificence of Solomon. To these gardeners, and I suppose they have always made up the majority, flowers represented an escape from the drab routine of earning a living and, as such, flowers could not be too grand, colourful, sumptuous or magnificent; the more they were removed from the ordinary, the more highly they were to be valued. As we have seen, the flower-gardeners of the Ancient World and those of the Middle Ages, had a small list to choose from—a few Roses and Lilies, Wallflowers, Violets, Pinks and Periwinkle—Snapdragons, Columbine, an Iris or two, and the Opium Poppy pretty well made up the total.

The "curious" plantsman

Two events were to transform the situation though, so far as concerns garden flowers, perhaps not so radically as is sometimes supposed. The first event was the discovery of the New World; the second, and more important, the emergence of a new race of gardeners, sometimes trained gardeners, but more often successful merchants, physicians, or professional men, well-educated and interested in the new learning, for whom the 16th century label, the "curious" is singularly apt and embracing. Some, like John

Gerard and John Parkinson, have been made famous by their publications, but most remain nameless, or like William Coys and Nicholas Lete, achieve nominal immortality in the pages of their great contemporaries. The line begins in the 16th century and continues through the Tradescants, Bishop Compton, Peter Collinson, Philip Miller, Canon Ellacombe and E. A. Bowles right up to the present day. We do not call them "curious" any longer, but I suppose "plantsmen" means the same thing, for their distinctive characteristic is an almost obsessional desire to acquire and cultivate novel plants, preferably species not seen in cultivation before, though, at a pinch, unlikely hybrids and remarkable cultivars will serve—in fact anything is grist for their mill, so long as it is not well known, or already to be seen in the gardens of their friends and rivals. They have none of that natural and primitive aversion to the wild, unmodified flower, but, on the contrary, glory in the prodigality of Nature, despising only that which time has honoured and artifice improved—the popular stock-in-trade of the nurseryman and seed-merchant. A weed, however undistinguished, will find a refuge in the plantsman's garden, if it comes fresh from Patagonia, and is not likely to be seen elsewhere. Indeed it would be no exaggeration to say that I have seen, in the collections of such enthusiasts, herbs which would bring nothing but discredit to any other garden.

Please do not think that I am belittling the activities of the plantsmen—on the contrary, the history of plant introduction shows that, at least up to the beginning of the nineteenth century, it was largely through the initiative of these men that we acquired, for the first time, such favourites as the Tulip, the Sunflower, the Nasturtium, the African and French Marigolds, the Sweet Pea, and many hundreds of the lesser known exotics which now adorn our gardens. It was through their enthusiasm that we derive the modern concept of a botanic garden and it was in answer to their persistent requests that travellers, sea-captains, settlers and ambassadors were obliged to become seed-collectors, and to look around for novel plant introductions. Yet, if we are indebted to the "curious" for many hundreds of the garden flowers we know today, it must at the same time be conceded that they rarely played a prominent part in the subsequent development of these garden flowers, that is, in the selection and breeding of cultivars which represent departures from the normal form of the wild, ancestral species. That activity was to become the province of a humbler class of gardener, the florist, and as the florist's enthusiasm waxed, the plantsman's enthusiasm waned. Admittedly, in default of other novelties, the gardens of Gerard, Parkinson and their successors were not without fancy sports and deviants from nature, but I doubt if these man-made flowers often originated in such gardens, nor do I think they were shown the same respect, nor so carefully cherished, as the strange novelties newly imported from beyond the seas.

The florists' flowers

The florists were not slow in making use of the newly introduced material; as early as 1587 Holinshed could record: "......in comparison with this present, the ancient gardens were but dunghills and laistowes to such as did possess them. How art also helpeth nature in the daily colouring, doubling and enlarging the proportion of our flowers, it is incredible to report; for so curious and cunning are our gardeners now in these days that they presume to do in manner what they list with nature and moderate her course in things as if they were her superiors.

It is a world also to see how many strange herbs, plants and annual fruits are daily brought to us from the Indies, Americas, Taprobane, Canary Islands and all parts of the world."

Holinshed (or more accurately William Harrison) correctly distinguishes the "cunning" gardener from the "curious" importers of strange plants: to the latter almost any novelty was welcome, whereas the former concentrated his energies on the selection and development of a relatively small group of plants which, in time, become known as "florists' flowers". Some of these, like the Carnation, were old-established inmates of the garden, others, like the Polyanthus, were partly derived from indigenous species, but the remainder, the Auricula, Hyacinth, Tulip, *Anemone coronaria* and *Ranunculus asiaticus* were recent arrivals from abroad. The "curious" were moved by a passionate desire to know more about nature and the world; the "cunning" were moved by a no less passionate, but distinctly more primitive desire to improve upon nature, and "to moderate her course in all things as if they were her superiors". The success of the florist's achievements in this directions is little short of astonishing: the ancestor of the garden Auricula is not likely to have reached north-western Europe much before 1580, yet within 80 years Sir Thomas Hanmer was able to list forty named sorts; the Tulip arrived about the same time, yet 17 years later, in 1597, Gerard could say that there were so many garden varieties that "to describe them particularly were to roll *Sisyphus*' stone, or number the sands". The efforts of the florist did not always make such speedy progress but, within their limited range of operation, the rate of development of garden cultivars, between 1580 and the mid-18th century, is little short of remarkable, and all the more remarkable for the fact that the exponents of this craft knew virtually nothing about the reproductive biology of the plants they propagated. The functions of stamens and pistils, and the mysteries of pollination and fertilization were virtually unknown until Camerarius published his *De Sexu Plantarum Epistola* in 1694, and the fundamental principles of hybridization and hybrid segregation did not begin to be understood until a hundred years later in the experiments of Thomas Andrew Knight, William Herbert, C. F. von Gaertner and others, culminating in Mendel's celebrated paper in 1865, which, as we know, was almost

completely ignored until the beginning of the present century.
Without any grasp of what we would now regard as the essential
basis of plant-breeding, one might wonder how it was that the
florists of two hundred, three hundred, indeed four hundred years
ago accomplished so much, and often in so short a time. Their
only equipment was a keen eye for promising variations, and their
only method was selection and skilful cultivation. It may well be
that their very restricted choice of subjects suitable for improvement
was not entirely accidental. At least two of the florists' flowers,
namely *Anemone coronaria* and *Ranunculus asiaticus*, exhibit even in
the wild an extraordinary capacity for colour variation and, like
allied species in the *Ranunculaceae,* have also a propensity for
"doubling", an abnormality which must have been an additional
recommendation to the florist. Of the remaining florists' flowers,
three, the Tulip, Polyanthus and Auricula, are generally supposed
to be of hybrid origin and would, in the circumstances, share that
instability which we have come to associate with hybrids. The
Carnation, whether we restrict the appellation to *Dianthus caryophyllus,* or extend it to include *Dianthus plumarius,* had a long history
of garden cultivation and development even before the age of the
florists began. It belongs moreover to a genus in which hybridization is rife, and indeed may have already been of mixed parentage
at the beginning of the 16th century if the yellow variety, reported
by Gerard, did in fact share some of its chromosomes with those of the
Balkan *Dianthus knappii.* Only the Hyacinth remains, presumably
of unsullied origin, but significantly less versatile than the others,
both in colour and form. On the whole, I think we may conclude
that the florists were not altogether arbitrary in their choice of subjects; nor should we overlook the fact that, with the Tulip, the
Ranunculus and the Anemone, a long period, perhaps centuries, of
development in Eastern gardens preceded the introduction of the
plant into Europe. Perhaps failure to achieve successfull results
in other directions may explain why the florists were so conservative,
and why Carnation, Auricula, Polyanthus, Anemone, Ranunculus,
Tulip and Hyacinth continued to reign supreme amongst domesti-
cated flowers up to the end of the 18th century. Thereafter,
systematic hybridization, and an increasing awareness of the scientific
principles of plant-breeding, begin to influence and direct the
development of garden flowers. To this new age of hybridization
we owe the cultivated Fuchsias, Astilbes, Delphiniums, Michaelmas
Daisies, Irises, Waterlilies, Aquilegias, Gladioli, Verbenas and
Rhododendrons, and indeed the vast majority of the developed and
transmogrified flowers that decorate our modern gardens and parks.
To the new age also must be assigned the Rose and the Pansy, for
though both of these have been part of the garden scene since time
immemorial, both have been so altered and transformed within the
past 150 years that they may be said to have entered upon a new
phase of existence. Another ancient garden flower, the Lily, has

indeed undergone a similar transformation within living memory, even within the memory of the comparatively young.

To attribute all these more recent developments in garden flowers to systematic breeding programmes and the hybridist would, of course, be an overstatement of the case. The modern Sweet Pea and the large-flowered Persian Cyclamen spring to fame through sportive extravaganzas on the part of nature, and very frequently, as with the Dahlia, Chrysanthemum and Lupin, extensive developments were made simply by rogueing and selection, the age-old techniques of the traditional florist. Even so, it cannot be denied that a background knowledge of plant reproductive mechanisms has, at least for the past hundred years, influenced, facilitated and very often accelerated the processes of contrived floriculture.

Was it a happy historical accident, or was it no accident at all, that this sudden outburst of flower-manufacture in the 19th century should have coincided with the great age of steam? The immense technological advances of the period we call the Industrial Revolution may have brought hardship to some; they also brought sudden affluence to a class who, a generation earlier, could scarcely have hoped for more than a modicum of independence as artisans, petty tradesmen and tenant farmers. This class had the natural and primitive tastes of their fathers, but in a much bigger, bolder and more flamboyant way. Not for them the muted rusticity of a Capability Brown landscape and the leisurely, contemplative ways of the old gentry. Trees take time to grow, and, anyway, it was the engine, and not Nature, that had made them what they were. So the gardens of their newly acquired villas were devised on mechanical principles, symmetry, rigidity and organization, plenty of cast iron, and carpet-beds as round and regular as a fly-wheel, and as bright as paint. Create the demand, and sure enough you will find a supply to meet it. In no time came Pelargoniums, Petunias, Zinnias, Dahlias, Verbenas, Begonias, Calceolarias, *Salvia splendens* and *Phlox drummondii*, gay as soldiers in their scarlet and gold, and almost as well disciplined—to the mechanical mind, divine; to the plantsmen, commonplace; to the gentry so pretentious that the mere mention of such plants still effects a perceptible nose-wrinkling amongst the upper echelons of the Royal Horticultural Society. It could not last, and it did not last. The cities grew larger, uglier and dirtier, and industry was found to create as many, and maybe more, problems than it solved. The sons were better educated, and consequently less buoyant, than their parents, and the bright, mechanical carpet-bedding began to lose its lustre. William Morris looked back nostalgically to the Middle Ages, and the less thoroughly romantic to a more recent never-never land of thatched cottages, roses round the door, old rustic bridges, bent apple boughs and contented peasantry. We had moved from success to sentiment, from the Palm House to the pastel informalities of William Robinson and Gertrude Jekyll. It was not a move back to Nature, but rather to a vision of Nature based upon a profound misunderstanding of the outlook and ambitions of those natural and primitive minds that had vanished in a puff of steam.

We still linger in the garden world of William Robinson; but it is near closing time. Aircraft have murdered peace; the woods decay and fall; the great garden hangs heavy about the neck of its owner, the smaller one languishes for lack of help. Soon the developers will move in, and fortunate indeed will be those who can boast three pots and a patio. Am I unduly gloomy? Perhaps. The age of the flower-developer is not yet over; there are still some elegant variations to be played on Mother-in-law's Tongue.

DISCUSSION

Dr W. T. STEARN remarked that as he had been referred to, he wished to say that Taprobane was Ceylon. As to a laistone, if Mr Meikle would care to speak to him later he would tell him then what it was.

WILD FLOWERS IN THE GARDEN

W. K. ASLET

Royal Horticultural Society's Garden, Wisley

A garden of native flowers

Ever since my Sunday morning walks in early childhood looking with my mother for the first Celandines, Violets, Primroses and Anemones, I have grown a few garden-worthy native plants. At one time it was one of my ambitions to plant a garden, or part of a garden, entirely with natives and, as a few of my friends may still remember, I did grow quite a considerable number. Others have, of course, done this sort of thing, and I remember that some years ago a Parks Department not far from London was trying to collect and grow *every* British species of flowering plant! (Lists of desiderata for this were sent around, and I recall causing indignation by absolutely refusing to pass on specimens of the Creeping Yellow Cress, *Rorippa sylvestris,* because I considered it the No. 1 enemy amongst our weeds. Every time I see this in a garden or nursery I do my best to persuade the owners to eliminate it at all costs).

Woody plants

From a garden point of view, out British flora has had much to offer, and these offerings, such as the Primrose, have been much too freely taken—to the detriment of our countryside today. Our gardens would, however, be very poorly off if we did become 'racialists' and eliminated all foreigners. We should have very few evergreen trees and shrubs, including only three conifers—Pine, Juniper and Yew—and three other evergreen trees—Holly, Arbutus and Box. We could have our specimen trees of Oak, Beech, Hornbeam and Birch, with Willow and Hazel for early catkins, Whitebeam for foliage and Mountain Ash for berries. Hedges, too we should have—Yew, Box, Beech, Hornbeam and Hawthorn— but not the ubiquitous privet, as our native one is very inferior for the purpose. We should be badly off for climbers—Ivy and Honeysuckle being the main woody ones, Old Man's Beard is hardly a garden plant itself but is still used sometimes as a grafting stock for Clematis cultivars. Our wild rose, too, is used in this way (although the Burnet Rose and Sweet Briar have produced some notable garden cultivars). For shrubs, *Viburnum opulus* has conspicuous flowers and fruits, and *Euonymus europaeus* brilliant fruits; both have good foliage colour in autumn. *Potentilla fruticosa* is a useful, smaller shrub for summer, and *Daphne mezereum* the earliest of all. The native Brooms, large and small, are all good and much used; even *Genista anglica* is an excellent garden plant.

Herbaceous plants

In the herbaceous border we could still have the Lythrums and a Thalictrum. Hemp Agrimony is used, but the colourful Ragwort would be a pest. *Achillea millefolium* and *A. ptarmica* are still numbered among the regulars. Even rock gardens benefit from our numerous mountain plants like *Saxifraga oppositifolia* and *Dryas octopetala,* but many lowlanders are at home here too, such as *Thymus drucei, Gentiana pneumonanthe* and the lovely *Pulsatilla vulgaris.* I still maintain at my place of work a stone sink-garden full of some of the tiniest natives.

Perhaps the section of the garden with the highest proportion of British species is the water garden and its surrounding boggy area. The Marsh Marigold, *Caltha palustris,* is in some ways the finest of all water loving plants, and *Hottonia palustris* (to many, an unexpected member of the Primrose family, the *Primulaceae*) is one of the most interesting. We see with this, and with the Bog Bean, *Menyanthes trifoliata,* and others, what a wide range of plant families have members adapted to an aquatic life. We even have bulbs for boggy places, like our native Fritillary and the Loddon Lily, *Leucojum aestivum,* whose bulbs sometimes grow quite happily completely submerged in water.

For shady places and ground cover we could still grow Aquilegias and Lily-of-the-Valley, Periwinkles and Pulmonarias, Yellow Archangel and that delightful little plant, *Lysimachia nemorum,* while the Bluebells, Solomon's Seals and Wood Anemones are not to be scorned, and there are many lovely ferns.

The herb garden has its Mints, Sweet Cicely, Chamomile and Chives. Vegetables like carrots, cabbages, beet and parsnips are referable to native species, and one parsnip cultivar, 'The Student', is reputed to have been deliberately bred in recent times from wild stock.

Some of our fruits, notably the raspberry, are true natives, but the cultivated strawberries are not.

Survival

In one way, of course, gardening is a fight to the death *against* our native flora, and in spite of modern techniques we can still find our special plants overwhelmed by "weeds"; the old countryman's saying, "One year's seed means seven years' weeds," is still a relevant tribute to their persistence, as we heard yesterday from Dr Chancellor. In demonstrations to students I often give them a list of invasive plants to beware of—especially as some of these are still freely sold by nurserymen!

The gardener's fight against Nature is largely due to his function of preserving a large number of Nature's aberrations or freaks that could not possibly survive in the wild. Pygmy conifers like *Juniperus communis echiniformis* and the witches' brooms in pine trees usually, like many double flowers, do not seed.

I go along wholeheartedly with our President, David McClintock, when he says that special variants should be propagated and preserved, if only for future study. Our heather gardens, for instance, would certainly be much the poorer if a few cuttings had not been taken, quite harmlessly, from the originals of such notable cultivars as *Calluna vulgaris* 'H. E. Beale'.

MINTS

R. M. Harley

Royal Botanic Gardens, Kew

The species

John Gerard in his Herbal of 1597 tells us that "the smell of mint rejoiceth the heart of Man", perhaps appropriately, as later this session we are to hear accounts of more festive plants such as snowdrops and mistletoe. Perhaps a more prophetic interpretation of his words may be found in the use of modern taxonomic methods such as gas chromatography to analyse the aromatic mint oils. These have provided welcome new characters to assist the taxonomist in his classification.

Another of Gerard's remarks seems equally appropriate. "Mint", we are told, "is a cure for the biting of mad dogs". In view of the plethora of worthless species and varieties described during the 19th and early part of the 20th centuries, with resulting taxonomic chaos, we can only conclude that certain taxonomists, who will be unnamed, took up the study of mints as a curative measure, unfortunately without success.

More than any other genus of comparable size, *Mentha,* poses problems of identification and classification. In Britain we have at most four native species. One of these, *Mentha pulegium* L., the Pennyroyal, causes no serious problems of identification. The remaining three are *M. aquatica* L. the Watermint, *M. arvensis* L. the Cornmint, and *M. suaveolens* Ehrh. the Applemint (which last has been incorrectly known by some as *M. rotundifolia*). The last of these three is also occasionally cultivated, and a variegated form exists.

In addition, there are two naturalized species, *M. requienii* Benth. from Corsica and Sardinia, and *M. spicata* L. the Spearmint. Only the latter causes taxonomic problems, and is widespread in cultivation and as an escape. It is the most important species in a consideration of introduced and cultivated mints. Both Spearmint and its numerous hybrids are widely grown and some forms are cultivated commerically for the essential oils.

The causes of complexity

In the short time at my disposal I will discuss the general causes of complexity of the genus, with special reference to *M. spicata.* The causes can be summarized under five headings as follows:—

1. *Morphological plasticity.* In common with many plants of damp habitats, the general appearance can be greatly altered by environmental factors. This can cause problems of identification which I do not intend to discuss here.

2. *Hybridization.* Mint flowers, being unspecialized and massed together, are visited by a range of insects, particularly flies.

There seem to be few incompatibility barriers to interspecific crossing, and hybrids are of frequent occurrence. Indeed, mint hybrids are more frequent than any others in the British flora. Though hybrids are usually highly sterile, some seed is set in many instances, and therefore back-crossing and segregation can occur. Except for *M. requienii* and *M. pulegium,* all species are capable of crossing, and even a triple hybrid, *M.* × *smithiana* R. A. Graham is frequent as a single sterile clone. *Mentha spicata* itself is an allotetraploid derived by hybridization and chromosome doubling from two diploid species.

3. *Vegetative propagation.* The vigorous rhizome system in mints allows clonal multiplication. In this way even sterile hybrids can persist and be dispersed. Some well known cultivated and naturalized taxa are single clones, e.g. *M.* × *villosa* Huds. nm. *alopecuroides* (Hull) Briq., often wrongly called *M. rotundifolia.*

4. *Cultivation.* There is a long history of cultivation in Europe, N. Africa and Asia, going back into antiquity. Cultivated mints are known from funeral wreaths from Thebes in Upper Egypt. Some extant cultivars have a venerable history, and their origins are obscure. Some cultivated mints easily become naturalized or can cross with wild species to produce a series of spontaneous hybrids which blur the distinction between what is native or introduced.

5. *Taxonomic confusion and ineptitude.* There are only ten species of *Mentha* recognized from Europe now, but in the *Index Kewensis* the names of over 900 species and hybrids are listed, chiefly from Europe. The majority of these are worthless, but it would be more than a life's work to reduce them to order.

Breeding behaviour and natural variation

The only logical way to study relationships in the genus and to elucidate the origins of some of the cultivated forms, is to start by trying to understand the breeding behaviour and patterns of variation in natural populations. By using cytogenetical and biochemical, as well as morphological information, we can hope to disentangle the more complex variation patterns found in the cultivated groups.

There is now firm evidence to indicate that *M. spicata* is a tetraploid species with 48 chromosomes derived from two wild diploids, each with 2n= 24, by hybridization and chromosome doubling. The diploids are *M. suaveolens,* a lowland species of the Mediterranean and Western Europe, and *M. longifolia,* a montane species widely distributed through Europe, N. Africa and Asia, but absent from Britain and most of NW. Europe.

These two diploids are ecologically separated, but can hybridize to produce fertile hybrids. In many cases, the fertility of the allotetraploid is in inverse proportion to the fertility of the diploid hybrid from which it originated. In fact, *M. spicata* shows some loss of fertility, particularly in some individuals. Irregularities at meiosis, and mis-pairing can result in offspring which revert towards one or other of the parental species in general appearance. In the past these have been treated as hybrids between *M. spicata* and the diploid which they most resemble. They are, however, fertile tetraploids and must be referred to *M. spicata.*

Nevertheless, genuine hybrids between *M. spicata* and *M. suaveolens* or *M. longifolia* frequently occur and are often cultivated. In some cases these are morphologically indistinguishable from forms of *M. spicata* such as those I have just mentioned, but unlike them, they are triploid and sterile. The complex as a whole cannot be satisfactorily treated taxonomically, and identification may be very difficult, though certain clones (or cultivars) are often easy to recognize.

An additional complication to confuse the issue further, is that although the garden (and therefore selected) forms of *M. spicata* are most often glabrous, hairy forms occur and are frequently present among progeny of *M. spicata* grown from seed.

In the past, these hairy forms of *M. spicata* have been referred to *M. longifolia,* a diploid which, however, is morphologically distinguishable. I have found, in breeding experiments, that hairiness is a recessive character controlled by a single gene, also transmitted to hybrids where *M. spicata* is one of the parents.

The classification of the hybrids has therefore been much confused, with hairy hybrids having been considered as derived from *M. longifolia* and glabrous hybrids from *M. spicata*. In breeding experiments, of course, it is possible to produce both types in the progeny of a single cross. A chromosome count will, however, indicate which of the two species is actually involved.

The selection of glabrous plants in cultivation probably arises from the fact that they possess a larger concentration of epidermal glands which contain the mint oils. However, the seedling progeny involving *M. spicata* which becomes established in the wild is frequently hairy, and it seems that hairiness has a higher survival value under such conditions. It does mean, also, that spontaneous hybrids in the wild often look rather different from their counterparts that have been subjected to selection under cultivation.

The two most important hybrid groups are *M.* × *piperita* L. (*M. aquatica* × *M. spicata*) the Peppermint, and *M.* × *gentilis* L. (*M. arvensis* × *spicata*). Both are important commercially outside Britain, and both occur here as garden escapes and spontaneous hybrids.

Finally, I would like to stress that many of these better known garden escapes are single clones, and, as has been suggested elsewhere, are best treated as cultivars. However, in a few cases we have no evidence of cultivation, and it would therefore be misleading to treat them in this way. Instead we must use the now accepted term nothomorph, that is for those forms of hybrid origin. However, to achieve uniformity in the group I can see no reason why a dual system of categories for these taxa should not be used.

PLATE I

Drosanthemum floribundum at New Grimsby, Tresco (above).
Oscularia deltoides on a wall at New Grimsby, Tresco (below).

Photos. by J. E. Lousley.

PLATE II

Galanthus nivalis 'Flore Pleno' in a copse by the River Avon, Wiltshire.

Photo. by R. D. Nutt.

PLATE IV

Galanthus nivalis ('Wisid; spires' at Thornworth, Dove, Sheffield

Photo. by R. D. Nutt

Galanthus nivalis 'Sandhill Gate' at Thornsett, Dore, Sheffield.

Photo. by R. D. Nutt.

PLATE VI

PLATE VII

Arum italicum × *maculatum*, from Jersey, cultivated at Farleigh, Surrey.

Photo. C. T. Prime.

Plate VIII

'WILD' GALANTHUS IN THE BRITISH ISLES

RICHARD NUTT

The status of the snowdrop in Britain

Are snowdrops wild in Great Britain or were they introduced and became naturalized, or are they just garden escapes? This is a vast subject; one needs go and look at nearly every patch of 'wild' snowdrops in the country! One must decide what species there are; one needs to understand fully what the variations are within a species. I will stick out my neck to say that as yet no one has studied the genus sufficiently well in the wild for one to be able to reach any conclusion, in spite of seven monographs. As far as I know, only one of them is based on extensive field work, that by Z. T. Artyushenko (1966), but she did not describe the limits of any species which she studied in the wild.

After such outrageous remarks, I will briefly try, with examples, to illustrate the problems and appeal to you all to go and look at areas of 'wild' snowdrops in Great Britain and let me know what you find.

In common with, for example, roses or daffodils, you will all recognize a snowdrop but, unlike the rose or the daffodil, how many of you know of any snowdrop other than the common single or double *Galanthus nivalis*? Added to this, I have many friends who say there are only two species and not about 17 as suggested by Artyushenko or about 12 by F. C. Stern (1956). My friends may well be right. The more I study the genus, the more I find I agree with them.

Stern proposed three series based on the differences in the vernation of the leaves. Other methods have been used and I have invented a few myself. However, from a horticultural viewpoint, and I am on the gardener's side of this hall, Stern's system has everything to commend itself.

The series are:—

> *Nivales* with the leaf vernation applanate or flat.
>
> *Plicati* with the margins of the leaves reduplicate or bent back.
>
> *Latifolii* with the leaves convolute or wrapped round each other.

Examples are known of members of each of the series naturalized in Great Britain. The only species which might be indigenous is *G. nivalis*.

Nearly invariably, one finds that snowdrops spread by vegetative reproduction rather than by seed. In the garden, be it wild or cultivated, I have found odd single flowering bulbs appearing, quite some distance away from the main mass. Let me add that my snowdrops, to a certain extent, have to be regimented or there would be a great muddle with my clones; I do not allow odd snowdrops to remain except in selected areas. Also, seedlings only occur with me adjacent to the parent, and that happens nothing like as frequently as in such gardens as Maidwell Hall or Highdown. From August onwards, when the bulbs are starting to grow, one frequently finds bulbs pushed onto the surface. These are then blown or carried away and when they find a suitable resting place, put out roots and pull themselves into the ground. Opportunity has not allowed me to observe, over a period of years, if this in fact happens outside gardens. But so far I have found only one place in the wild where the leaves of seedling are to be found. As far as I am aware, the wild double form of *G. nivalis* only spreads by offsets, and in many places the doubles are more extensive than the singles.

What happens outside Great Britain? As far as is known, only one species of snowdrop is sterile, *G. bortkewitschianus*. It occurs in a very small area of the Caucasus. I have seen wild snowdrops in the Pyrenees, Southern Dolomites, Central Italy and the Lebanon. Except for Central Italy, when I was too late to observe the leaves properly, those in all the other areas clearly reproduced themselves by seeding. One found a mass of seedling leaves where the scape had laid the seed capsule on the ground. The snowdrops were in pockets—not in the white masses which we have. When digging up the bulbs they were singles, rather than clumps as in our own *G. nivalis*.

At one time it was thought by some galanthophiles that the common *G. nivalis* never set seed in this country. Several people have disproved this and last July I collected seed from *G. nivalis*. It will be interesting to see if it germinates. Of the 'wild' snowdrops, which I have seen, mainly in Norfolk and Wiltshire, only at one location in a wood by the River Avon at Rushall, just south of Pewsey, was there definitely the single leaf of seedlings. It was noticeable that there were no doubles at this site, while fifty yards away, on the other side of the road in a pollarded willow copse, the plants were predominantly double with some singles. In my view, these doubles were obviously not garden escapes as are the doubles, a mile down stream, at Netheravon. Incidentally, it was from Netheravon that Margery Fish received her 'Pewsey Green'. I have been sent double snowdrops from there by her old friend Maud Drew, but there has never been any sign of green on the outer segment green tips. At Netheravon there is a large military camp and 'wild' singles and doubles grow in great profusion. It may be that at the Hall there are double "Green Tips".

Reverting to the question of whether snowdrops are indigenous, naturalized or just garden escapes, some guide might be expected to be found in various general and local floras. Bentham & Hooker, 7th. edition (1924:470), seemed to think that it was 'probably not indigenous, but long cultivated, and now naturalized in England, Scotland and Ireland'. Clapham, Tutin & Warburg in their second edition (1962: 999) opine to 'Probably native', and continue '. . . . but very commonly planted and usually naturalized'.

On the assumption that snowdrops were introduced, the questions that need to be answered are when did they come to Britain and from where? It has been suggested that they were brought over by the Romans to remind them of home. There is a snowdrop which for want of a better name we might call *G. imperati*. It is wild near Rome and can be found near Naples and other parts of Italy. It does not appear quite identical to our own *G. nivalis*. It is said that the Romans grew snowdrops in the hill camps in this country, but I have been unable to obtain confirmation of any sites. If they did, they were much better cultivators than we are. A. H. Church (1908: 17) says "there is little doubt that . . . . the Snowdrop was introduced from the Mediterranean during the centuries of Roman occupation", but quotes no authority. His detailed observations on the snowdrop have never been equalled, as far as I am aware.

Perusal of the first mention of snowdrops in Great Britain and reliable identification requires a detailed study of the early herbals, not least those printed on the Continent. Briefly Gerard in *The Herbal* (1597: 120) has *Leucoium Bulbosum Praecox* the "Timely flowering Bulbus violet" and the description, but not the illustration, which is of a convolute snowdrop, would appear to fit *Galanthus nivalis*. Gerard says the snowdrop is wild in Italy and common in London gardens.

I had expected to find a reference in William Salmon's *Botanologia, The English Herbal* (1710) but failed to so do. Perhaps someone has spotted it there. Miller in the first edition of *The Gardeners Dictionary* (1731) says that the snowdrop is cultivated and not wild. This would appear to conflict with Church's suggestion that the Romans introduced the bulbs.

From my observations one usually finds wild snowdrops in calcareous areas, be it the chalk round Cambridge, the soils of Norfolk or the red soils of Berkshire and Northumberland. Snowdrops seem to prefer the shade of deciduous woods or the partial shade of banks and churchyards; I have yet to hear of their being sunlovers. A high water table is also the rule, yet there are famous sights of steep banks covered with snowdrops, but I feel that these are the exception. These conditions are vastly different to the open, dry hill tops where the Romans made their camps.

There appear to be two guides as to what may have happened to the Roman's snowdrops, if, in fact, they introduced them. The obvious pointer comes from other genera which have been commercialized by the Dutch, who have been incredibly successful in selecting bulbous plants, be they *Crocus* or *Scilla,* which are beautiful in form, like being grown in sandy soil (albeit with a highish water table), multiply rapidly by producing offsets and do not object to being taken out of the soil and dried annually.   It is because snow-drops do not like being lifted that the Dutch are not successful with them.   Less obvious guidance is obtained from collected bulbs, and here I can only really speak about snowdrops.   There are four types, those which just die out as they do not like us, those which multiply but rarely flower, those which flower and set viable seed but do not increase vegetatively and, best of all, those which flower and occasionally set viable seed but multiply vegetatively.   I suggest that at sometime or other the last type was introduced to Great Britain, for which we must be grateful.

It occurs to me that collected bulbs might behave differently in cultivation in other and hotter countries.   They might set seed but not produce offsets.   Has anyone any information on this?

At several of the sites of wild snowdrops I have visited, I was told that they were planted at such and such a date; quite often at the end of the last century.   In other places it is only too clear that they are garden escapes.   Under what conditions do the snowdrops naturalize themselves?   As I have said, a high water table and calcareous soils seem to be a general requirement. together with half shade or deciduous woods, even though the ground may be, and frequently is, covered with ivy.   (I mention the latter as Lewis Palmer could never grow snowdrops in the ivy at his graden on chalk at Headbourne Grange, near Winchester, although they seeded all over where there was no ivy).   I would not expect in this country to find snowdrops in full sun, nor in heath land, nor in sandy soils, and so far have not so done.   I remember driving near Andover on one of the Romans' straight roads and stopping suddenly at the sight of snowdrops in the hedge.   It was in chalk country with the water table miles down.   Earlier in the day I had looked in many likely woods with no success, so these few bulbs were doubly welcome. Quite clearly they were wild.   Other snowdrops in hedgerows have always had a source in the adjacent park or garden.   Clearly snowdrops like the conditions in the hedges: winter damp, half shade and dryness in the summer.

The conditions produced in churchyards nearly invariably suit snowdrops, but here the bulbs have spread from being planted on a grave, and clearly they have naturalized themselves.   Whereas they are lovers of deciduous woods, which provide sun in the spring when the flowers come out, in the churchyards the snowdrops grow under yews in what one would have thought to be too deep a shade.   I know of one churchyard where *G. elwesii* is doing its best to take possession.

The above remarks apply to both single and double snowdrops. Sometimes I have found only singles, or only doubles, but generally both together with one form predominating.   Very frequently in Norfolk (five out of twelve sites) *Eranthis hyemalis* is present and those of you who have, on a spring sunny day, seen the combination of snowdrops and Winter Aconites will, I hope, agree that nature has painted a most beautiful picture for us. At eight sites in Berkshire and Wiltshire there was no *Eranthis*. The gardeners amongst us will know that the Winter Aconite is not easy to please; it either likes you and multiplies, or sulks and just exists. That it seems to require more exacting conditions than *G. nivalis* is not surprising if you accept it as being an introduction. At each of the five sites in Norfolk *Eranthis* was clearly introduced.

Natural forms of Galanthus nivalis

Alas, I have not been able to study these in the field to the extent I should wish, but the following are known.

Galanthus nivalis 'Flore Pleno' (Plate III). When and where did these plants originate? Is there an earlier record than Miller in *The Gardeners Dictionary* (1731) under "3. *Narcisso-Leucojum; trifolium, minus, flore pleno. Boerh. Ind.* The Double Snow-drop", which is "preferr'd to the Single, for the Largeness and Fairness of its double Flowers".

Names other than 'Flore Pleno' have been used: 'Plenus', but by whom and when I cannot say, and 'Hortensis' by Herbert (1837: 330).

There are said to be different forms, some more regular than others. Both James Allen (1891) and F. W. Burbidge (1891) at the Snowdrop Meeting organized by the Royal Horticultural Society in 1891, referred to two or three separate forms, some being more regular than others. It is not too difficult in the garden but on paper I find it hard to separate them out and define the differences. Certainly, in some years, due to the weather no doubt, and in some soils, one finds double snowdrops infinitely more irregular. The best example of a regular flower is a cultivar, *G. caucasicus* 'Double'. With me, in Sheffield on heavyish soil, very rarely indeed are any of the inner segments aberrant (or larger than each other and mis-shapen). Until this year I would have said never. With most people I find that this snowdrop has occasionally aberrant inner segments, but this year they were very frequent indeed. Was it the dry autumn?

Mrs Joan Reynolds, who is writing, a thesis on double flowers writes to me, "you are so right, people fight shy of any direct reference to the origin of almost all double flowers. I find this a most interesting study but cannot pretend that I am anywhere near to solving it . . . . even the nature of doubleness is very variable and the first reference is often casual".

A number of writers have referred to the single snowdrop giving place to the double. F. W. Burbidge (1891) drew attention to this and suggested that it might be a fungoid disease common to the single snowdrop, or lack of moisture or some other reason. The Rev. Joseph Jacob (1924: 51–52) says ' . . . . the bulbs, if they are left alone, will soon take to producing double flowers in place of the normal single'. Earlier, D. T. Fish (1884 : 6–10) said he had difficulty in growing the single snowdrop for any length of time in several places, and that "the single snowdrop is constantly running into the double forms". He also makes the point that snowdrops increase by offsets—seldom from seed. This year I heard of an area in Hampshire where double snowdrops were predominating at the expense of the singles but have not been able as yet to visit them. My own experience in Sheffield, where I grow both singles and doubles in grass, the soil being medium clay, is that the singles multiply at least twice as fast as the doubles.

Galanthus nivalis 'Pusey Green Tips'. This is a double snow-drop which is found wild round Faringdon in Berkshire. The outer segments have fused green lines at their tips. There appear to be two forms, even in the same area, one in which the outer segments are of reasonable width and not clutching the inner segments, and one in which the inner segments are narrower and clutch the inner segments, almost like a claw. This latter form often has split spathes as in *G. nivalis* 'Scharlokii', a natural bulb from the River Nahe in Germany. The split spathe of 'Pusey Green Tips' is certainly not constant. One can find patches of the ordinary double adjacent to 'Pusey Green Tips', and I should give a word of warning that many of the double snowdrops in the churchyards around Faringdon do not have green tips. Visiting a garden of a galanthophile in Painswick, Gloucestershire, a double *G. nivalis* with green tips was pointed out to me. It had naturalized itself many years before.

Single G. nivalis with Green Tips. There was a time when I thought that green tips on the outer segment were very rare but I now know of two lots of *G. nivalis* with them and I am quite certain there are others. (I should add that in other species there are occasional flowers with green tips which may or may not be constant). My first Green Tips were found in 1969 near Faringdon in a churchyard—there was one clump only. The second lot was in a rectory garden about five miles west of Cambridge where there are single and double *nivalis*, both of which have spread, and some of the single snowdrops have green tips. I saw them in the rectory this year, but let me add that the owner had already spotted them. Both these lots, like 'Pusey Green Tips', keep their green tips in other gardens.

Galanthus nivalis 'Viridi-apice' *(Plate IV), while not known wild in Britain, should be mentioned as Mr Oliver E. P. Wyatt found a very similar snowdrop in Northamptonshire. He has

---

*The spelling as first published in the *J. Roy. Hort. Soc.* **48**: xxxi (1923) was 'Viridapicis', and Viridiapicis' has also been used, but 'Viridi-apice' is better Latin,—Ed.

called it 'Courteenhall' after the place. As it only came to me in 1970 I have yet to determine how it differs from 'Viridi-apice'. This latter was exhibited to the Royal Horticultural Society Scientific Committee in 1922 by Messrs Barr, who had obtained it from Van Tubergen, in whose nursery in Holland it had appeared (in which country other apparently similar Snowdrops were also known). With me it is a bigger thing than G. nivalis when grown under the same conditions. The outer segments have at the apex five green lines, which are usually separate and not fused together. The spathe is much larger, both longer and wider than in G. nivalis, and can be split, but next year it will be united. This is not the place to go into the merits or validity of G. nivalis 'Warei', a very large spathed form of 'Viridi-apice'.

Very small forms of G. nivalis occur in Great Britain; some have been found in Norfolk but I have no record of where. There is one under the name of 'Tiny Tim', which was put out by Sir William Lawrence. I know nothing more about it, except that it is still in cultivation.

Galanthus nivalis 'Poculiformis' and its relations should be characterized by having the inner segments lengthened to about the same length as the outer, and little or no green marking on the apex of the inner segments, which have only two green lines on the inner face as opposed to the usual four pairs. I say should as it is a variable plant which does not always behave itself; again it seems to be a question of soil and climate. 'Poculiformis' was a name used by H. Harper Crewe (1880), who relates that the snowdrop was found among seedlings by D. Melville, gardener to the Duke of Sutherland at Dunrobin Castle. Whether it qualifies for the term "wild" in the context of Great Britain may be doubtful but I mention it to encourage people to go and look.

James Allen at the 1891 Snowdrop Meeting relates, as does F. W. Burbidge, of a similar snowdrop being found in Wales by A. D. Webster, and, in addition, how he had received bulbs of a very similar form from a lady near Ayr, in whose garden it grew. A. D. Webster lived in Llandegai and the woods at Penrhyn had exciting forms of snowdrops. William B. Boyd of Faldonside, Melrose, when addressing the Berwickshire Naturalists' Club in 1905 relates that the finest form of this snowdrop was from Sir George Douglas, found in the woods at Springwood Park. He also had other bulbs from Mrs Grey Milfield, from her garden. I have seen a self sown seedling in a garden in Somerset but the moles got in before it could be increased. In 1968 the Rev. Richard J. Blakeway-Phillips sent me a very similar form which he calls 'Sandhill Gate' (Plate V); it was found by him in a cottage garden in Crawley Down. It has never behaved well with him but I always have all white inner segments with no green markings, the inner ones are just shorter than the outer. The orange stamens give it a faint orange glow when seen from the outside.

If you are not convinced that this is a common snowdrop in Great Britain, my friends on the Continent tell me that it grows in sheets under the name of *G. nivalis* 'Hololeucus' and is still known by that name. I now have bulbs from Vienna and await their flowering to see if there are any differences.

The most intriguing form of *G. nivalis* is called 'Lutescens' or 'Flavescens' or 'Howick Yellow'. I would prefer not to be embroiled in nomenclature and priorities of date of publication here. I was in Northumberland near Belford in February this year and saw over quite an area of wild *G. nivalis*, both single and double. Here and there would be a single bulb or a clump of a snowdrop with yellow ovary and yellow markings on both sides of the inner segments, where they are normally green. The scape could be yellowish too. They did not appear to be seedlings; I say 'appear' as it was appallingly cold and wet. The amount and shade of yellow appears to vary and I have a friend who is collecting the different forms. The soil is rich, reddish loam and would appear to be acid. The yellow colour seems to remain constant but seedlings have been recorded as being either yellow or green.

Lastly, there is *G. nivalis* 'Lady Elphinstone' which originated at Heawood Hall in Cheshire in 1890, and I do not claim this was wild but it may have been. The green markings on the inner segment are yellow, some years, in some soils. I suspect there are other double yellows waiting to be found.

Other naturalized species

Alas, I know of only two other species which have become clearly naturalized in Great Britain.

Galanthus elwesii. The limits of variation in this Snowdrop in the wild are not known—in fact, I know of only one lot of 'collected' material in cultivation. However, a number of clones are grown, each of which arose from one bulb and is now spreading vegetatively (as I have suggested happened with *G. nivalis*), so they should be mentioned.

Galanthus elwesii 'Lanarth' is possibly the best known. It is a clone which flourished at Michael Williams's garden in Cornwall and was a white sheet as seen by E. A. Bowles. It has distinct apical and basal markings.

Galanthus elwesii 'Ladham's Variety' is less well known, and I have not been able to lay to earth what is definitely the true plant. Mr James Platt tells me that he remembers two groups of *G. elwesii* under the dark sycamore in Ladham's garden at his Nursery at

Elstead, Godalming, Surrey. There was a large and a small form
and hundreds of bulbs of each. It was in 1937 that Ladham brought
up the bulbs to a Royal Horticultural Society Show. Mr Platt
obtained bulbs and distributed them. Does anyone know whether
this variety is still in existence?

I know of at least one garden in Berkshire where *G. elwesii*
has spread by seeding itself under a cut-leaved beech. The evidence
is that the bulbs were planted at about the turn of the century and
that they liked the situation, set seed and spread. Double flowers, and
stems with two pedicels have been seen. Incidentally, this year I
was sent two lots of snowdrops with two pedicels, *G. elwesii* ex 'the
stores' and *G. byzantinus* which had appeared in a hedge near
Stourport in Worcestershire.

Again, near Faringdon, along a hazel walk and in the copse
behind, I saw between 10,000 and 20,000 bulbs of *G. elwesii* believed
to have been first planted there about sixty years ago. There
were seedlings all over. The green inner segment markings varied
from apical and basal to all green, or to apical only. *G. caucasicus*
appeared to be there. There was one clear hybrid with *G. nivalis,*
having glaucous applanate leaves, wider and taller than this latter
species, and the flowers had its typical apical marking, be it a
good strong mark.

Near Bury, there is what appears to be *G. nivalis* × *plicatus*
naturalized, but I have not personally seen the plants.

Galanthus 'Straffon' came from Ireland. I am quite clear in
my own mind what it looks like, but some of the story of its origin
must be apocryphal. Further, there is a snowdrop called 'The
O'Mahoney' which would appear to be identical or at least those
bulbs so labelled which I have seen. E. A. Bowles in his chapter
VII in Stern's *Snowdrops and Snowflakes* (p. 68) admits to the touch
of mystery about it and does not seek to solve this. This snowdrop
flowers after *G. nivalis* has reached its peak. The glaucous leaves
have very slight plication, usually on one margin only, so that at
first examination it can be taken for a form of *G. nivalis,* but the
flowers are white and larger, added to which when established it
will regularly put up two scapes. I have yet to hear of *G. nivalis*
so doing.

There are two accounts of it—the first at the Snowdrop Meeting
by F. W. Burbidge (1891) who, under the name *G. grandis,* says it
is "a fine form of *G. nivalis* of the *caucasicus* type"; a statement I
fail to understand as *G. caucasicus* has convolute leaves. He went on
to say that "This form came to the gardens at Straffon, co. Kildare
along with *G. plicatus,* being brought by Lord Clarina on his return
from the Crimea". James Allen in his paper at the same meeting,
without providing a name, refers to Lord Clarina in 1856 bringing
home with him from the Crimea some snowdrop bulbs which he
had collected whilst there, and goes on to refer to the 'Straffon
Snowdrop'. Phyllis, Lady Moore, (1959) relates that Lord Clarina

sent bulbs of *G. plicatus* from the Valley of Tchernoya home to his sister, the Hon Mrs Barton of Straffon. In 1858 Mr Bedford, the head gardener noticed the different flowers. Somewhere else I found a note that within a very few years there were thousands of bulbs. Anyhow in the meadow by the river Anna Liffey at Straffon, this snowdrop has naturalized and was still flourishing in 1959.

Snowdrops probably not native

If the definition of snowdrops being wild in Great Britain depends on their spreading by seed, they are most certainly not wild from the evidence which I have been able to muster. If the definition of snowdrops being wild in Great Britain admits that they spread only by being able to divide freely and through some form of invertebrate distribution of the bulbs, then they are wild; but I am not inclded to accept this definition.

I did not set out to solve any problems but possibly I have inadvertently so done. If any of you know of wild snowdrops, please let me know so that I can come and see them .

REFERENCES

ALLEN, J. (1891). Snowdrops. *J. Roy. Hort. Soc.* **13**: 172–188.
ARTYUSHENKO, Z. T. (1966). Taxonomy of the Genus *Galanthus* L. *R.H.S. Daffodil & Tulip Year Book* **32**: 62–82.
BENTHAM, G. & HOOKER, J. D. (1924). *Handbook of the British Flora.* 7th Ed. London.
BOYD, W. B. (1905). Annual address covering Snowdrops in cultivation. *Proc. Berwicksh. Nat. Club* **19**: 233–249.
BURBIDGE, F. W. (1891). Snowdrops. *J. Roy. Hort. Soc.* **13**: 191–210.
CHURCH, A. H. (1908). *Types of Floral Mechanism.* Oxford.
CLAPHAM, A. R., TUTIN, T. G. & WARBURG, E. F. (1962). *Flora of the British Isles.* 2nd Edition. Cambridge.
CREWE, H. H. (1880). *Galanthus nivalis* 'Poculiformis'. *The Garden* **17**: 249.
FISH, D. T. (1884). *Bulbs and Bulb Culture.* London.
HERBERT, W. (1837). *Amaryllidaceae.* London.
JACOB, J. (1924). *Hardy Bulbs for Amateurs.* London.
MILLER, P. (1731). *The Gardeners Dictionary.* London.
MOORE, PHYLLIS, LADY. (1959). *Gard. Chron.* III, **145**: 225.
NUTT, R. (1968). Some thoughts on growing Snowdrops. *R.H.S. Daffodil & Tulip Year Book* **34**: 80–86.
STERN, F. C. (1956). *Snowdrops and Snowflakes.* London.

MISTLETOE

FRANKLYN PERRING

*Biological Records Centre, Monks Wood Experimental Station,
Huntingdon*

Introduction

In the winter of 1969/70 the Botanical Society of the British Isles launched a survey of the distribution of Mistletoe in the British Isles. It was thought that it was one of the few species which could be mapped accurately on the basis of its presence in the 2×2 km squares of the National Grid for the whole country, and that at the same time it would be of value to find out the range of hosts on which it occurred and the relative abundance of the species on different hosts.

After three winter seasons (it is difficult to see Mistletoe after the end of May) most of the country has been covered, as a result of the efforts of members of the Society, and the general public which has responded to publicity on radio, television and in the daily press.

Overall, it is clear that the map published in the *Atlas of the British Flora* (Fig. 1) underestimated the frequency of the species, particularly in the north and west. There are many more records in Yorkshire, Durham, Cumberland and Northumberland than was previously appreciated and here I would like to pay particular tribute to Mr F. Stubbs, who with the assistance of the Northern Horticultural Society organised an enormous survey in the north of England, starting with a letter in the *Northern Gardener*. The species occurs more frequently in Scotland too. There was but one record in the *Atlas* but the survey has added another eight and though, according to Miss E. Beattie, it was introduced into Old Dean Cemetery, Edinburgh, in about 1863 by William Paxton, it still flourishes and has now spread to four crab apples, three hawthorns and two limes.

Distribution and range of host species

The survey returns soon began to show that the distribution pattern revealed is very much modified by the direct and indirect activities of gardeners and the owners of estates, past and present. Directly because, certainly at the limits of distribution in the north and widely elsewhere, Mistletoe only survives where it has been deliberately planted, usually after Christmas on an apple tree in the garden. Indirectly, because, as the analysis in Table 1 shows, by far the greater percentage of Mistletoe growing "wild" in Britain is on non-native or planted trees, and the amount growing on native trees in native situations is negligible.

Mistletoe has been recorded on 62 species and varieties during the survey, of which 45 are planted trees. Of the first five species in order of frequency as hosts of Mistletoe in 10 km squares, four are aliens, which are widely planted, and the fifth *Crataegus monogyna* is, in the majority of its sites, of planted origin (See Figs. 2 to 6).

The list is strong in *Acer* (7), *Populus* (8), *Malus* (6) and *Prunus* (4), the last excluding a yet unknown number of ornamentals, but it also includes some highly exotic species confined to Botanic Gardens or the gardens of large houses, e.g. *Davidia involucrata, Cladrastis lutea* and even *Vitis vinifera* (in a greenhouse).

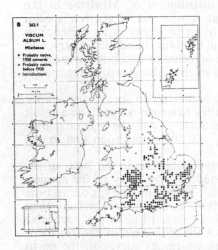

FIGURE 1
Distribution of Mistletoe from *Atlas of the British Flora* (1962).

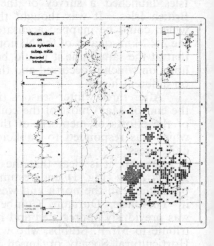

FIGURE 2
Distribution of Mistletoe on *Malus sylvestris* subsp. *mitis*.

The list of native hosts is just as interesting for the species not included or on which Mistletoe is very rare, as for what is included. The most notable absentees are Alder, Beech, Sessile Oak and Gean and the native Conifers, but there are only single records of Holly, Hornbeam and White Willow. Noteworthy too is the great rarity of Mistletoe's occurrence on Birch (3), Elm (5) and Pedunculate Oak (12), three of the most widespread trees in the landscape, especially when it is remembered that when I refer to 10 km squares these records invariably represent only single trees. I am also suspicious that the numbers of Elms and Oaks are inflated by mis-identifications. Some field botanists who would not have the slightest difficulty of separating Elm and Lime in summer apparently find them a problem in winter and most Elm records I have challenged have been withdrawn. Oak has, I think, sometimes been mistaken for *Robinia* which is a more widespread species than certainly I had appreciated. It could perhaps be identified as the large tree with zig-zag branches and Mistletoe. It is unlikely that there are more than a dozen oaks with Mistletoe now remaining in Britain and most of these are very old, dying trees. Clearly the religious significance of mistletoe on oak must have been in part due to the rarity of its occurrence.

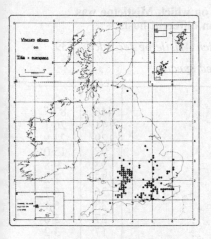

FIGURE 3
Distribution of Mistletoe on
Tilia × europaea

FIGURE 4
Distribution of Mistletoe on
Crataegus monogyna

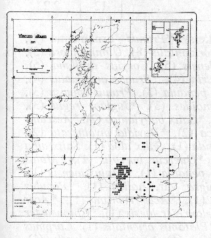

FIGURE 5
Distribution of Mistletoe on
Populus × canadensis

FIGURE 6
Distribution of Mistletoe on
Robinia pseudoacacia.

We are then left with a very short list of native species which make a significant contribution as hosts for Mistletoe Hawthorn (129), Field Maple (39), Crab Apple (36), Crack Willow (33), Ash (27) and Rowan (15), though several records for the last named are from town trees. One is therefore bound to speculate on

TABLE 1

Numbers of 10 km squares on which Mistletoe was recorded growing on different hosts

Total number of squares in which Mistletoe recorded=539

*=Introduced species

		%
Malus sylvestris subsp. *mitis*	446	82·7
Tilia × *europaea*	141	26·1
Crataegus monogyna	129	23·9
Populus × *canadensis*	97	15·0
Robinia pseudoacacia	53	9·8
Acer campestre	39	7·2
Malus sylvestris subsp. *sylvestris*	36	6·7
Salix fragilis	33	6·1
Fraxinus excelsior	27	5·0
Acer pseudoplatanus	17	3·2
Pyrus communis	17	3·2
Sorbus aucuparia	15	2·8
Prunus domestica	13	2·4
Aesculus hippocastanum	12	2·2
Quercus robur	12	2·2

Also (numbers of squares in brackets): *Prunus*—ornamental (9), *Populus tremula* (8), *Cotoneaster horizontalis* (8), *Ulmus glabra* (5), *Populus nigra* (5), *Populus gileadensis* (5), *Corylus avellana* (5), *Mespilus germanicus* (4), *Sorbus aria* (4), *Prunus amygdalus* (4), *Laburnum anagyroides* (3), *Betula pendula* (3), *Prunus spinosa* (3), *Acer palmatum* (3), *Acer platanoides* (3), *Juglans regia* (3), *Acer saccharinum* (3), *Rosa* spp. (2), *Salix babylonica* (2), *Aesculus octandra* (2), *Chaenomeles japonica* (2), *Davidia involucrata* (2), *Populus alba* (2), *Wisteria floribunda* (1), *Salix alba* (1), *Salix alba* ×*fragilis* (1), *Ilex aquifolium* (1), *Quercus borealis* (1), *Malus lemoinei* (1), *Malus prunifolia* var. *pendula* (1), *Malus ioensis* (1), *Malus purpurea* (1), *Populus nigra* var. *italica* (1), *Populus deltoides* (1), *Populus trichoides* (1), *Acer rubrum* (1), *Acer ginnala* (1), *Tamarix* sp. (1), *Amelanchier laevis* (1), *Cotoneaster lindleyi* (1), *Cladrastis lutea* (1), *Platanus orientalis* (1), *Carpinus betulus* (1), *Philadelphus coronarius* (1), *Syringa vulgaris* (1), *Vitis vinifera* (1), *Crataegus orientalis* (1), *Sorbus commixta* (1), *Aesculus* × *neglecta* (1).

where, if at all, Mistletoe was a native species before the advent of man—or at least horticultural man.   Two observations are helpful, one is that these are mainly small trees (not woodland dominants), the second is that Mistletoe hardly ever occurs on trees in woods: it is something nearly always seen on isolated trees, on hedges or avenues. Therefore, as a native, I think it can only have occurred in naturally open light woodland: such a place I believe was the Wye Valley in Monmouth and Hereford where Mistletoe occurs on Ash and Whitebeam, on the steep cliffs along the river above Tintern.

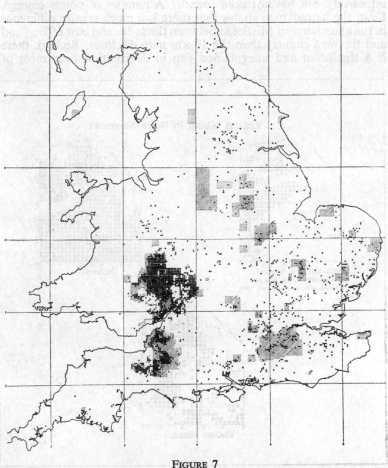

FIGURE 7

Distribution in England and Wales of 2×2km. squares in which Mistletoe has been recorded during the survey.

Distribution

Whilst it is now extremely difficult to assess the native distribution of Mistletoe, it must be one of the easiest species of which to map the actual distribution. It can be identified with certainty at distances of a mile and more across open countryside and, when hidden in gardens, information gleaned from the 'local' or the Post Office can normally be relied on as there are no problems of identification.

The details, showing the 2×2 km squares in which Mistletoe was recorded during the survey, are given in Fig. 7. Here the hatched lines indicate areas where Mistletoe has been searched for intensively, but has not been found. A number of points emerge. First, the 'tetrad' map shows that there is a much greater difference in the abundance of Mistletoe between the south and east of England and the west country than the 10 km map discloses. Second, there is a significant and unexplained gap in distribution over most of

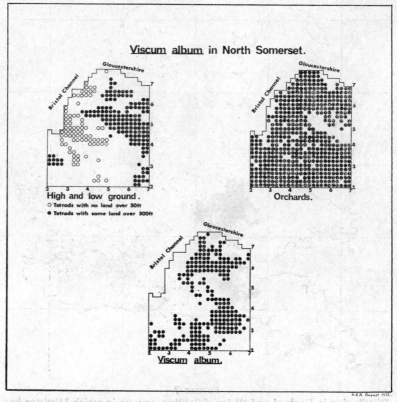

FIGURE 8
Distribution of Mistletoe, orchards, and high and low ground in the N. Somerset area.

Wiltshire, Oxfordshire and the southern half of Northamptonshire. Lastly, that the survey is still incomplete. Meetings organized in Hereford and Somerset in Easter, 1971 and 1972 respectively make it clear that the boundaries have not yet been determined with complete accuracy and more remains to be done and will, I hope, be achieved in the winter of 1972/73 in south-west Somerset and north-east Hereford, south Shropshire and north-west Worcestershire.

Nevertheless, in the well worked areas of Somerset and Herefordshire it was found that the limits of distribution are extremely clearly marked. Within a mile it is possible to move from areas of abundant Mistletoe to areas with none at all. It might have been assumed that distribution depended upon the distribution of orchards— in both Somerset and Herefordshire cider apple orchards are abundant. But the survey showed that in both areas the distribution of orchards is more extensive than that of Mistletoe. Fig. 8 shows the distribution of Mistletoe in the north Somerset area, and the distribution of orchards taken from the 2½in. O.S. map. There are many thousands of acres of apple orchards in Somerset, and to a lesser extent in Herefordshire, adjacent to the areas where Mistletoe is abundant, which have no Mistletoe and appear never to have had any.

Altitude appears to be a far more significant limiting factor. Fig. 8 also shows the distribution in north Somerset of tetrads with land over 500ft. and similarly tetrads with no land above 50 ft. Mistletoe is largely absent from both these classes of tetrad but frequent in most of the others. An upper altitude limit is not surprising, but whereas it is about 500 ft. in Somerset, it was at about 900 ft. in Herefordshire, and it was noticed that the maximum altitude at which Mistletoe occurred rose as the height of the terrain rose, but was always 50–100 ft. below the maximum altitude of the square. It does not grow on exposed plateaux but may occur on the slopes below, as long as these do not exceed 900 ft.

The lower altitude limit in Somerset is very surprising but may be related in some way to exposure to the influence of winds from the Atlantic sweeping up the Bristol Channel—the higher up the Channel the nearer it grows to the coast.

Further speculation would not be helpful—but an investigation into the growth and reproduction of Mistletoe along the limits of its distribution, and artificial introduction beyond that limit could be most rewarding. Now that the surveying phase of Mistletoe study is nearing completion I hope that Botanical Society members and gardeners who live near the line might assist in the experimental phase.

GARDEN ESCAPES AND NATURALIZED PLANTS

C. M. ROB

Plant introductions

Over the years plants have been introduced into this country
either by accident or for some particular purpose; some of these
have escaped and now form part of the flora, competing with the
other plants of the district and generally behaving as natives—
although in fact, some of these escapes at times overpower, rather
than compete with established species.

Plants have been introduced for many reasons, some, which
have come by accident, as crop impurities, grain aliens, with bird
seed or in wool, rarely persist, although a few, such as *Veronica
persica,* are now common weeds of arable land and waste ground.

Medical plants or pot herbs

Plants were introduced for medicinal purposes, or as pot-herbs.
Good King Henry, *Chenopodium bonus-henricus,* has been with us
for many years, but even now grows, as a rule, around farms, on
roadsides and village greens, relict of the time when it was used
as a vegetable; either the leaves were boiled like Spinach or the
young shoots cooked as one would Asparagus. Although the
plant is common over much of England it is rare in Western Wales
and Scotland. Alexanders, *Smyrnium olusatrum,* another pot-
herb, persists near some ruined castles and Abbeys often near the
sea, where is has escaped on to the surrounding cliffs.

Garden escapes

Many of the most striking escaped plants are those which were
grown in gardens for themselves alone, either for beauty or for
interest, and it is from these I have chosen a small number to
talk about, plants which I know well in the North, most of them in
North Yorkshire.

In the past, until in fact work on producing the *Atlas of the
British Flora* was started, garden escapes were all too often looked
upon as something rather improper, either not looked at, or if
noticed dismissed as "just another escape", as though they were
something not very nice. One reason for this was the difficulty of
finding the correct name for any non-British plant in the books
available; there have in the past been records of rare native plants
in quite impossible places, just because some hopeful botanist tried
to fit a plant to something in his particular flora.

Because of this failure to appreciate the importance of some of
the introduced plants, a few species became established with little

or no notice taken of them, making it difficult if not impossible to trace the history of their spread and colonization in Britain.

One example of this is Himalayan Balsam, *Impatiens glandulifera*, also known as Policeman's Helmet. This plant came from India in 1839 as a garden plant, escaping onto river banks nearby it soon spread and sixteen years later was well established; by 1932 it was recorded in 27 vice-counties, thirty years later it was in 47 and the spread is still going on. Very abundant in the industrial north it grows quite happily by rivers thick with detergent foam, colonizing ground where few plants could survive. The explosive fruit can throw the seeds (some 8–10 per capsule) for a considerable distance, making it as easy for the plant to work upstream as down; although generally still a plant of the lowland areas, this tall pink-flowered alien is working its way steadily up most of the Pennine dales. It is rather less invasive in North East Yorkshire, possibly one reason being the lower rain fall, nevertheless here the plant has left riversides and is now in wet places in the hills. In some four years, with its high rate of germination, the offspring of a single uninvited plant in the drive of my house, now number several thousand, some reaching a height of ten feet, nearly as tall as my next plant the Giant Hogweed.

This huge plant, *Heracleum mantegazzianum,* introduced from the Caucasus in about 1839, has come in for a great deal of publicity in recent years. No one was interested until 1943 when reports of the plant causing skin trouble appeared, but for over twenty years no one seemed very concerned. Recently, particularly on Teeside, there has been considerable alarm after some children suffered from contact with the Hogweed, the percentage of people sensitive to this plant (a condition accelerated by sunshine) is small, the majority being able to handle it with no ill effects. There has been a lot of confusion as to the exact identity of the species causing the trouble, first Hemlock was blamed, but once the real culprit was named, pressure was put on to have this giant newcomer destroyed, the cry went up "let us Spray". Contrary to the general theory, Giant Hogweeds have been in or near the Teeside area since 1911 to my personal knowledge. This is truly a giant plant with stems up to 3·5 m and a stem like a small tree. In addition to Teeside the plant is abundant in many parts, particularly around London and in Northern Ireland.

Slender Speedwell, *Veronica filiformis,* a prostrate creeping plant was originally introduced as suitable for rockeries. Spreading vegatatively it is now a troublesome weed of lawns and other parts of gardens; frequent in church yards and on roadsides, it is well on the way to becoming a really bad weed of damper grassland. Generally a wash down from gardens it can spread by flood over riverside fields, forming a dense mat which inhibits the growth of grass, so spoiling the quality of the grazing. Like many invasive weeds it tends to settle down after a year or so, all the same it is difficult to get rid of once it comes, the Stray at Harrogate is a sheet of blue in spring, as are some of the pastures by the Eden in Westmoreland. Weed killers can and do help, but they add considerably to the cost of farming.

New Zealand Willow-herb, *Epilobium nerterioides,* was recorded as a weed in a garden at Craigmillar near Edinburgh in 1904, four years later it came with shrubs to Elmet Hall and Calverley near Leeds; by 1930 it was across the Vale of York, in the Esk valley, at Egton Bridge and other places around Whitby. Since then this tiny fragile creeping plant with pink flowers and long slender fruits has turned up in many places; some extremely remote, miles from habitation or even a road. It prefers the wetter climate of the west and there appear to be few if any records from the south east. It grows on rocks and stones in mountain streams, by tracks across the hills, in old quarries, even clambering over *Sphagnum,* the small round leaves looking not unlike our native Bog Pimpernel. First recorded as *E. nummularifolium,* it was then realized that the plant was not this species but *E. pedunculare,* today the accepted name is *E. nerterioides,* although a name change is again about to happen. [It is now called *E. brunnescens,* see *Watsonia* 9: 140. 1972—Ed.] There is some doubt as to all the British plants being the same species, so again more change seems likely.

Several species of *Mimulus* have become well established. *M. guttatus,* native of North America, was first recorded as an escape at Downton, by the Avon. In 1860 F. A. Lees writes of it as "A spateborne stray in a garden garth at Arnecliffe". Now *Mimulus* is abundant, generally in the west. *M. luteus,* with, in spite of its name, more red on the flower, is very rare and only in the north, while the species with copper coloured flowers, *M. cupreus,* is never found established as an escape. There is a lot of confusion as to which species is which, the distribution is not yet known, and the situation is made worse because the majority of our plants are hybrids between these three species. *Mimulus moschatus,* once grown for the scent which has now gone, a hairy, flabby plant with much smaller flowers, is well established in wet places in many parts of Britain, although rare in Ireland. All species of *Mimulus* favour areas of high rainfall or where the water supply is constant, *M. moschatus* in particular is often found far from habitations.

Calystegia silvatica, Bellbind, looks at first sight very much like the native *C. sepium.* It is larger and has inflated bracteoles. The known distribution shows it to be more common in the north, but because of confusion between the two species the information may not be very accurate. *C. silvatica* has increased very much in the last twenty years. The reason for this is not known, possibly the practise of tipping waste soil from road improvements on to verges of remote roadsides may be one reason. I have been assured by one rural gardener that the birds carry it around. The story that it was introduced by an unscrupulous nurseryman as a plant "Suitable for odd corners, will grow anywhere" may or may not be true, it was certainly here as long ago as 1777.

Rhododendron ponticum is a weed in many forestry plantings, as well as being a menace in some of our interesting peatlands where it spreads at a rapid and alarming rate. It has been, in the past, planted as game cover and is, when in flower, an attractive sight, but as it is very difficult to control, even with modern herbicides, most foresters would rather not have it in their woodlands; but the ornithologists find it a useful plant for birdlife.

Pink Purslane, *Montia sibirica,* is widespread in many places, generally in the wetter parts of the north and west. This again is a plant with a story. It is said the plant was introduced into the Manchester area, particularly along the Ship Canal with cotton. I cannot vouch for the truth of the story, but in 1927 Pink Purslane was plentiful in old quarries between Manchester and Halifax, while in 1940 it appeared to be in every garden and on much waste land in and around Salford. It is often grown in gardens and the colony in Bransdale in NE Yorkshire is known to have escaped from a wild garden at the head of the dale. Once again this plant, in common with other escapes is generally under-recorded. The closely related *M. perfoliata,* Spanish Lettuce, Spring Beauty or Buttonhole Plant, is also found as a weed in many places, but has not spread in the vigorous way of Pink Purslane.

There are at least two species of *Cicerbita,* Blue Sow-thistle, known in Britain. The one generally recorded as *C. macrophylla* is a large untidy plant with lilac-blue flowers and greyish foliage. Little seems to have been published about this group, the first Yorkshire record in 1935 was in the East Riding. Some of the early records were given as *C. alpina,* which caused something of a sensation. It was soon realized that the plant was in fact a native of Southern Europe and the Caucasus, and since then it has been recorded from a number of localities.

Of the three species of Butterbur which occur as escapes, only Winter Heliotrope, *Petasites fragrans,* is really common, occuring by roadsides, on railway banks, in churchyards and similar places. *P. albus* with white flowers and *P. japonicus,* with large yellow green bracts and smelling of toothpaste, rarely get far from where they were planted, although they persist for many years.

Erinus alpinus, is a popular rockery plant which often escapes onto nearby walls, there are many records for it, particularly in the Glasgow area, but once again it is very under-recorded. It is in some old quarries in NW. Yorkshire, where it seems likely to have been deliberately introduced, as in fact it is at Housesteads on the Roman wall, in spite of the romantic story of its persistance since the Roman occupation.

The same story of persistance since long ago s told about *Dianthus plumarius* on the walls at Fountains Abbey, no one knows in this case if the monks were responsible for its introduction, but it is known that in recent years Pinks have been planted on the walls to help keep the population going. *D. plumarius* is found on other old buildings notably Beaulieu Abbey, while *D. caryophyllus,* the ancestor of the garden Carnation, according to McClintock, "Haunts the Gents lavatory in a certain old building in Kent".

One of the most striking and beautiful garden escapes is the Wild Lupin, *Lupinus nootkatensis,* from NW. America, which is so well established on some Scottish river shingle, surviving the violent spates for which these rivers are so well known. I know little about this plant except that it it beautiful, more so than Tree Lupin, *L. arboreus,* which is established on sand dunes in many places and on the embankments of some new motorways, though in this last case it may be planted.

These are a few of the many escaped and/or introduced plants in Britain today, there are many hundreds more.

DISCUSSION

Mr R. L. GULLIVER questioned the use of the phrase "rate of germination". He believed that if seeds germinated well one should say that one had a high percentage germination. If they germinated rapidly, then one had a high rate of germination.

Miss ROB replied that she had really meant a high percentage germination. An enormous number of young plants had come up, far more than she would ever have expected from the one plant that survived in her garden. The rate of spread was quite phenomenal.

Dr H. HEINE said that he had recently been to southern Scotland and went on an excursion with colleagues from the Edinburgh Botanic Garden. They saw the River Tweed's banks crowded with Giant Heracleums. He had been told that it was not only *H. mantegazzianum* that was there and thought that one of the very tall species had been identified as *H. persicum,* He understood that these introduced plants even hybridized with the native species although, unfortunately, he had not seen any of the hybrids.

Miss ROB replied that she had not gone in to the matter herself. She had heard that more than one species was involved but she certainly had not done any work on them. All the ones that they had in Yorkshire appeared to be the same species, for which she had used the name *H. mantegazzianum,* and she had not seen any other in their very small area that looked in any way different.

Mr D. McCLINTOCK said that there were in fact two species, or two taxa, of Giant Hogweeds. One was almost certainly *H. mantegazzianum.* He had had it in his garden for years and the tallest specimen there in 1972 was 17 feet high. The great character of *H. mantegazzianum* was the fact that it was monocarpic, which meant that for three or four years it built up to the flowering stage, flowered and then died. This character could never be shown on herbarium specimens, but one did not even find a note to this effect. It was only by growing it in his garden that he had discovered that this giant one was monocarpic. One could usually tell it by the stem for often there was a seedling by its side, making it look as though it had come up a second year. However, the rosette was really a seedling growing by the old stem.

He further commented that he also grew another species which was not quite so tall. It had leaves which were not quite so dissected and it was perennial—he had had some clumps in his garden for 12 to 15 years. The best place to see this species was in the middle of the old part of the Oxford Botanic Garden where there was a large plant. It was labelled *H. mantegazzianum,* but although it was not that species he could not immediately say which it was, but its name might be *H. lehmanniana.* To settle the question involved examining some type specimens from Russia and the matter was still being pursued. It had been known that there were these two taxa for some time. The flowers and fruit were somewhat similar, the leaf was different and the life-duration different. He did not even know whether they were species, subspecies or what, but they were quite distinctive taxa.

For some time he had been interested in the question of hybrids between *H. mantegazzianum* and the ordinary wild Hogweed, *H. sphondylium*. Various people were quite sure that they hybridized, and he had been sent to various colonies where it was thought there were hybrids, but, to his own satisfaction, he had neatly divided the plants as *H. mantegazzianum* or *H. sphondylium*, without hybrids. About two years ago in Edinburgh he had seen Miss Muirhead and discovered that quite independently she had been working on this problem and she convinced him where others had failed. He now believed there were hybrids and had learnt the look of them. They were quite hard to tell, but now he had even found them in his own garden.

LITERATURE OF PLANTS

R. DESMOND

Royal Botanic Gardens, Kew

Introduction

In the very short time at my disposal it is not possible for me to do more than to skim the surface of the enormous range of books and periodicals of potential value to both botanist and horticulturist. If I omit your favourite book I apologize most humbly.

In compiling my list I have assumed that you all have access to a reasonably good library since a number of the books I have chosen are either unfortunately out of print or prohibitively expensive to buy second-hand.

Plants names

Botanical and horticultural books require frequent revision because of necessary changes in botanical nomenclature. It was John Ruskin, I believe, who ridiculed the use of scientific names in preference to the more homely vernacular ones. He had an ally in that great Victorian gardener, William Robinson, who argued that "old English books like Gerard were rich in English names, and we should follow their ways and be ashamed to use for things in the garden a strange tongue—dog Latin or as it may be". But however paradoxical it may seem to the gardener, the use of scientific names does introduce stability and the assurance of reliable identity into the naming of plants. David McClintock's *A Guide to the naming of plants, with special reference to heathers,* published by the Heather Society in 1969, sets out to explain to gardeners and nurserymen in non-technical language the principles upon which the correct naming of plants is based.

In November 1951 an international committee met at the Royal Horticultural Society to consider the nomenclature of cultivated plants. The well-established *International Code of Botanical Nomenclature* provided the guide-lines for the formulation of the *International Code for the Nomenclature of Cultivated Plants* (1953) which sought "to promote uniformity, accuracy and fixity in the naming of agricultural, horticultural and silvicultural cultivars". This Code is kept up to date by frequent revisions, the latest being in 1969.

The *Index Kewensis,* which owes its genesis to a suggestion of Charles Darwin, lists all validly published generic and specific names of flowering plants. The first volume appeared in 1893 and additional new volumes are recorded in quinquennial supplements.

The 7th edition of J. C. Willis's *A Dictionary of the flowering plants and ferns,* edited by H. K. Airy Shaw (Cambridge University Press, 1966) includes every generic name (whether validly published or not) from the publication of Linnaeus's *Species plantarum* (1753) and all family names from Jussieu's *Genera plantarum* (1789). Unfortunately, this new edition excludes the definitions of botanical terms which were such a useful feature of previous editions of Willis. This omission makes B. D. Jackson's *A Glossary of botanic terms with their derivation and accent* indispensable. Edition 4 (1928) which was reprinted in the 1960s is still available.

The second edition of A. W. Smith's *A Gardener's dictionary of plant names* (Cassell, 1972) has been extensively revised by Dr W. T. Stearn. It lists nearly 6,000 botanical names, their origin, meaning and pronounciation, followed by an alphabetical index of 3,000 vernacular names with their scientific equivalents. It is a fascinating book to dip into. Did you, for instance, know that *Euphorbia robbiae* was named after Mrs Robb of Liphook in Hampshire who brought it back to this country in her hat-box— hence its name 'Mrs Robb's Bonnet'.

Indexes and encyclopaedias

Most of you must have heard at some time that venerable radio programme, Desert Island Discs, where the imaginary castaway is always asked to select one book to comfort him in his solitude. If I were asked to recommend just one book to serve a gardener it would unhesitatingly be the Royal Horticultural Society's *Dictionary of gardening* (1951). Perhaps this is slightly cheating since it consists of four volumes and a second edition of the Supplement (1969). Its lineage extends back through George Nicholson's *Illustrated dictionary of gardening* (1884–1901) to revisions of Philip Millers's classic *Gardeners dictionary*. The many specialist contributors give this work its authority and reliability. For instance, the best account of ornamental grasses to be found anywhere is by Dr C. E. Hubbard in the revised Supplement.

The American counterpart of the *Dictionary of gardening* is L. H. Bailey's *Standard cyclopedia of horticulture,* first published in 1914 and subsequently revised and reprinted. Its American bias does not detract from its usefulness to British gardeners. It is well-equipped with keys to families and genera, and is generously illustrated. Supplementing it is L. H. and E. Z. Bailey's *Hortus Second* (Macmillan Co., 1941) which is "designed to account for all the species and botanical varieties of plants in cultivation in the continental U.S. and Canada in the decades ending mid-year 1940" (Preface). *Hortus third* has been in course of preparation for some years.

Plant illustrations

Right at the beginning of *Alice in Wonderland,* Lewis Carroll's young heroine petulantly observes: "What is the use of a book without pictures", a sentiment surely appreciated by most botanists and gardeners who frequently need pictures of plants for identification or inspiration. Consequently, there is a long tradition of illustrated botanical and horticultural books extending back to the primitive woodcuts of old herbals. The golden age of illustrated flower books began about the middle of the eighteenth century, and the next hundred years or so saw its consummation. This was the era of hand coloured engravings and lithographs in periodicals such as *Curtis's Botanical Magazine* (1787 onwards), *The Botanical Register* (1815–1847), *The Botanical Cabinet* (1817–1833), *The Botanic Garden* (1825–1851), Paxton's *Magazine of Botany* (1834–1849) and many more. Despite the flood of illustrated books today, these earlier works often provide the only adequate pictorial records of many cultivated plants.

With the aim of simplifying the tiresome task of finding specific plant illustrations the Royal Horticultural Society and the Royal Botanic Gardens at Kew joined forces to compile the *Index Londinensis* (6 vols., Clarendon Press, 1929–1931) which took G. A. Pritzel's *Iconum Botanicarum* (1855; Supplement, 1866) as its exemplar. Books and periodicals from the eighteenth century onwards were thoroughly combed for illustrations of flowering plants and ferns. A supplement in two volumes in 1941 brought the period of coverage up to 1935. The half a million or so bibliographical references in the *Index Londinensis* are arranged alphabetically by species within their respective genera. Illustrations which are in colour or are confined to flowers, fruit or habit are appropriately distinguished. It must be remembered that because no attempt was made to resolve synonymy in the *Index Londinensis*, the same plant can be recorded under a number of different names. A new supplement covering the years 1936 to 1960 is being compiled at Kew. From Supplement 10 (1936–1940) onwards the *Index Kewensis* has asterisked all descriptions accompanied by illustrations.

The *Index Londinensis* is fine for the person who has access to a large library, but most people would prefer a single volume which they can take down from their own bookshelves. A book that admirably fulfils this function is *The Dictionary of garden plants in colour with house and greenhouse plants* (Ebury Press & Michael Joseph, 1969), by Roy Hay and Patrick Synge. Sponsored by the Royal Horticultural Society to supplement their *Dictionary of gardening,* it has over 2000 coloured photographs grouped under conventional horticultural categories such as alpine and rock garden, annual and biennial, greenhouse and home, and so on. Published at a remarkably low price it is of outstanding value. With six times the number of photographs, mostly in black and white, A. B. Graf's *Exotica* 3 (Roehrs Co., 1963) is inevitably a substantial quarto volume. It sets out to be an exhaustive pictorial encyclopaedia of exotic indoor plants. *The Exotic plant manual* (1972) is a condensed version of *Exotica* 3.

Horticultural periodicals

On 2nd January, 1841 the first number of *The Gardener's Chronicle* appeared, a red-letter day in the history of British horticulture. The general editor, Joseph Paxton, delegated the botanical content of the paper to the distinguished botanist and industrious Assistant Secretary of the Royal Horticultural Society, John Lindley. No gardening periodical was every launched with a more impressive combination of talent and enterprise. Over the years it has had many distinguished contributors, such as the Kew taxonomists, J. G. Baker and N. E. Brown, who between them wrote an extensive series of articles on "New garden plants" from 1884 to 1897. Some of the best line-drawings of plants executed by that much denigrated botanical artist, Worthington G. Smith, graced its pages about the same period. The 230 volumes of the *Gardener's Chronicle* are a veritable treasure trove for botanist and gardener. Alas, the inadequate annual indexes make much of this information virtually inaccessible.

The same shortcoming cannot be ascribed to the *Journal of the Royal Horticultural Society* which, in addition to good annual indexes, has a series of cumulative indexes making retrospective searching back to the first volume in 1846 so delightfully easy. Essential reading for the dedicated gardener, its inclusion of important monographic revisions also makes it a necessary reference tool for the botanist. One can select almost at random papers such as R. Lloyd Praeger's *An Account of the genus Sedum as found in cultivation* (vol. 46 (1921); 1–314), W. Wright Smith and George Forrest's *"The Sections of the genus Primula* (vol. 54 (1929): 4–50), and H. W. Pugsley's *A Monograph of Narcissus, subgenus Ajax* (vol. 58 (1933): 17–93).

The announcement last year that the Royal Horticultural Society's other serial publications—*The Daffodil and Tulip year book, The Lily year book* and *The Rhododendron and Camellia year book—*were to cease publication was most disturbing news for all specialists working on these groups of plants. The R.H.S. Lily Committee is to be commended for so swiftly bringing out *Lilies* 1972 *and allied plants* as an interim successor to *The Lily year book*. I understand that the Royal Horticultural Society have now agreed to publish somewhat attenuated year books for Daffodils and Rhododendrons for an experimental period.

Within the narrower field of horticultural taxonomy the American quarterly periodical, *Baileya,* published by the New York College of Agriculture is pre-eminent. Since its inception in 1953 it has reviewed the cultivated species of families such as the *Compositae, Liliaceae, Primulaceae* and *Scrophulariaceae. Baileya* is named after L. H. Bailey who deserves the gratitude of garden-lovers everywhere, not only for his *Standard cyclopedia of horticulture* but also for his *Manual of cultivated plants* (Macmillan Co., 1949). The latter work has concise descriptions of nearly 32,000 species and varieties (excluding horticultural varieties); there are keys to families and genera; and every family is typified by line-drawings of respresentative plants.

Line-drawings, from four to six to a page, are also introduced into F. K. Makins' *Herbaceous garden flora* (Dent, 1957) to assist in the identification of about 1,000 species of herbaceous perennials, biennials and annuals cultivated in our gardens. J. W. C. Kirk's *A British garden flora* (Arnold, 1927) also has line-drawings of hardy and half-hardy plants. Both books are furnished with adequate keys.

Orchids

In the nineteenth century no group of plants generated more literature than the orchids. For rich and prosperous Victorians there was no status symbol more satisfying than an orchid collection which out-rivalled all others. Regarded as curiosities in the eighteenth century, it was not until the Loddiges nursery at Hackney began to import and cultivate orchids about 1812 that they were taken seriously. John Lindley was an early victim and wrote several notable books about them. Many of the opulently illustrated Victorian orchid books were intended for rich collectors. Among them were J. Bateman's magnificent *Orchidaceae of Mexico and Guatemala* (1837–1843) and F. Sander's *Reichenbachia* (1888–1894). Cheaper and much more informative for identification and cultivation purposes were B. S. Williams' *The Orchid-grower's manual* which reached its 7th edition in 1894, and James Veitch and Sons' *Manual of orchidaceous plants cultivated under glass in Great Britain* (2 vol., 1887–1894). The popularity of orchid cultivation in the closing decades of the nineteenth century created a demand for this relatively cheap comprehensive manual which combined sound practical advice on cultivation with good descriptions of species and varieties. The recent facsimile reprinting of Williams' book indicates that it has not yet lost its usefulness. Two post-war books, although dealing with orchids on a regional basis, are of interest to British orchid growers. D. S. Correll's *Native orchids of North America* (Chronica Botanica, 1956) has good descriptions of species with cultural notes by two specialists. A truly impeccable piece of scholarship is Professor R. E. Holttum's *Orchids of Malaya* (ed. 3, Govt. Printer, Singapore, 1964). His taxonomic descriptions include Asiatic and American species as well as Malayan. The *Orchid Review*, founded in 1893 by R. A. Rolfe of the Kew Herbarium, is now the oldest orchid periodical in existence.

Other monocotyledons

Continuing with the monocotyledons, C. H. Grey's *Hardy bulbs* (3 vols., Williams & Norgate, 1937–1938) surveys in some depth the genera of seven families. Since a second-hand copy now costs nearly £100 most gardeners must be content with Patrick Synge's *Collins guide to bulbs,* the 2nd edition of which appeared last year. Uniform with Collins' other flower guides in the same series, it is excellent value.

H. G. Elwes' wealth enabled the publication of one of the most impressive examples of Victorian botanical iconography, namely his *Monograph of the Genus Lilium* (1877–1880). The fluent line and panache of W. H. Fitch's splendid plates have not always met with unqualified approval. That fastidious gardener, E. A. Bowles, complained that Fitch has "a way of making a flower look artificial, as though made of painted calico and not of living sap-filled cells." I suspect he greeted more favourably the supplements published between 1933 and 1940 which were illustrated by Lilian Snelling. The plates to supplements 8 and 9 (1960 and 1962) were drawn by Miss M. Stones. Elwes' book was aimed at the botanical bibliophile collector whereas *Lilies of the world; their cultivation and classification* (Country Life, 1950) by H. B. Drysdale Woodcock and W. T. Stearn was written for the less affluent gardener. Part I covers cultivation, and Part II, the major part, describes *Lilium* species and hybrids.

F. W. Burbidge's *The Narcissus: its history and culture* (Reeve, 1875) was the first really significant study of this group of flowers. The 48 plates by the author illustrating all known species compensate with botanical accuracy what they lack in artistic competence. All E. A. Bowles' writing is permeated with his love and enthusiasm for plants; his *Handbook of Narcissus* (Hopkinson, 1934) is a light wine after Burbidge's strong ale. The book opens up with a typical Bowles aside. "If a motto were sought to be inscribed in large letters at the entrance to a Daffodil garden, could a better be found than the terse Arab proverb, 'Dig on a hundred days, water on one?'" Similar casual comments and personal reminiscences leaven the experience and knowledge which he brought to his *Handbook of Crocus and Colchicum* (ed. 2, Bodley Head, 1952), still the only modern study of these two genera. John Ruskin hailed George Maw's drawings for his *Monograph of the Genus Crocus* (1886) as "most exquisite.... and quite beyond criticism". Maw grew in his Shropshire garden most of the species which he described at great length, together with their history, geographical distribution and cultivation. A copy of this book was recently offered for sale at £370. I believe a continental publisher is now considering a facsimile reprint.

W. R. Dykes, a former Secretary of the Royal Horticultural Society, had two great loves—Irises and Tulips. His *magnum opus, The Genus Iris* (Cambridge University Press, 1913) is well-illustrated with good keys, lengthy descriptions, cultivation notes and a detailed bibliography of Iris literature extending back to pre-Linnean times. This is a book primarily for the botanist, whereas his *Handbook of garden Irises* (Hopkinson, 1924) is intended for the gardener. *Notes on Tulip species* (Jenkins, 1930) was edited and illustrated by his wife after his tragically early death. In 1940 the Royal Horticultural Society published Sir Daniel Hall's *Genus Tulipa* which, like Maw's work on the Crocus, was based on first-hand observation of the living plants over many years. The Royal

Horticultural Society was also responsible for another authoritative work: *Snowdrops and Snowflakes* (1956) by Sir Frederick Stern who originally started it in collaboration with E. A. Bowles. It is still the best available treatment of *Galanthus* and *Leucojum*. Miss Christabel Beck's *Fritillaries* (Faber, 1953) brings to a close this cursory look at the literature of the monocotyledons. In just under 100 pages she surveys the cultivated species of Europe, Asia and North America from the gardener's point of view.

Dicotyledonous plants

Miss Ellen Willmott's *Genus Rosa* (2 vols., Murray, 1910–1914), with coloured plates by Alfred Parsons, leads us into the dicotyledons. J. G. Baker's collaboration ensured the accuracy of the taxonomic descriptions. The Royal Horticultural Society sponsored the publication of Sir Frederick Stern's *Study of the genus Paeonia* (1946), G. H. Johnstone's *Asiatic Magnolias in cultivation* (1955) and J. R. Sealy's *Revision of the genus Camellia* (1958). All have achieved the status of standard works. For many people the name of Sir George Taylor is associated as much with his *Account of the genus Meconopsis* (Flora & Silva, 1934) as with his directorship of Kew Gardens. Sir George's botanical critique is complemented by cultivation notes by E. H. M. Cox on introduced species.

From time to time the publishers, "Country Life", have been responsible for some useful gardening books; for instance, H. C. Crook's *Campanulas* (1951) and David Wilkie's *Gentians* which reached a second edition in 1950. *Campanulas* is the first monographic treatment of the genus since Alphonse de Candolle's revision in his father's *Prodromus* in 1839. A thorough botanical revision still remains to be written but it can be recommended for its horticultural content.

Ericas have been garden favourites ever since Francis Masson, a Kew plant collector, began sending them back from the Cape in the late eighteenth century. Since James Andrew's well-known works on heaths about 150 years ago there had been no other illustrated work devoted exclusively to South African Ericas until 1967 when *Ericas in Southern Africa* (Purnell) by H. A. Baker and E. G. H. Oliver appeared. Altogether 167 species are described and illustrated with coloured drawings, but because these represent less than a third of the South African species no key has been provided.

H. C. D. de Wit's *Aquarium plants,* published in the Netherlands in 1957–1958, was followed by an English translation in 1964 (Blandford Press). This has now been joined by a German version which would seem to indicate some international recognition of this work. Essentially it is an amateur's book giving brief non-technical descriptions of aquatic plants interspersed with frequent cultural notes.

Cacti and succulents

Probably no group of plants has enjoyed more popularity in recent years than the succulents. New societies such as the African Succulent Plant Society (1966), the Mammillaria Society (1960) and the Sempervivum Society (1970) have been formed by enthusiasts. *The National Cactus and Succulent Journal* was started in 1946 by the National Cactus and Succulent Society. This would appear to overlap with the older *Cactus and Succulent Journal of Great Britain* founded in 1932 by the Cactus and Succulent Society of Great Britain. Both periodicals are written for the amateur collector and grower, although occasionally they carry articles of interest to the botanist. Far more relevant in a taxonomic context is the *Journal of the Cactus and Succulent Society of America* (1929 onwards), now titled the *Cactus and Succulent Journal.* Two other foreign periodicals deserving a mention are the German *Kakteen und andere Sukkulenten* (1937 onwards) and the Dutch *Succulenta* (1947 onwards), both monthly publications.

Unquestionably the best general guide to succulents is Hermann Jacobsen's *A Handbook of succulent plants; descriptions, synonyms and cultural details for succulents other than Cactaceae* (3 vols., Blandford Press, 1960). Translated from the German edition of 1954–1955 it has good concise descriptions accompanied by many black and white photographs. Short descriptions of genera and species can be found in Jacobsen's *Das sukkulenten Lexikon* (Fischer, 1970) which also includes over 1,000 small black and white photographs on 200 plates.

Mesembryanthema (Reeve, 1931) by N. E. Brown, A. Tischer and M. C. Karsten, for some years a standard work, is now superseded by H. Herre's *The Genera of the Mesembryanthemaceae* (Tafelberg-Uitgewers Beperk), published last year. Illustrated with 124 coloured drawings by artists of the Bolus Herbarium in Cape Town, this new study is intended for the botanist and grower. Also dealing with the same family is G. Schwantes' *Flowering stones and mid-day flowers, a book for plant and nature lovers on the Mesembryanthemaceae* (Benn, 1957) in an good translation by Vera Higgins.

Still in South Africa we have A. White, R. A. Dyer and B. L. Sloane's *The succulent Euphorbieae* (*Southern Africa*) (2 vols., Abbey Garden Press, 1941) which was compiled on the same lines as A. White and B. L. Sloane's *The Stapelieae* (3 vols., Passadena; 1937), a well-illustrated and exhaustive treatment of this tribe of the *Asclepiadaceae*. A second edition of G. W. Reynold's *The Aloes of South Africa* (Balkema), which omits most of the illustrations appearing in the first edition, came out in 1969 to accompany his *Aloes of Tropical Africa and Madagascar* (Aloes Book Fund, 1966). Neither work is likely to be superseded for many years.

A Gardener's guide to Sedums (1970) by R. L. Evans was the first of the Alpine Garden Society's excellent guides. The Royal Horticultural Society, which included R. L. Praeger's review of

Sedums in its *Journal* in 1921, also published in 1932 his *Account of the Sempervivum Group*.

In emulation of the *Index Kewensis* the *Repertorium Plantarum Succulentarum* has been recording annually since 1950 new botanical names and, latterly, adding a list of current literature.

Descriptions of 124 genera and 1,253 species, more than 1,200 illustrations, including 146 coloured drawings, frequent keys to genera, subgenera and species qualify *The Cactaceae* (4 vols., Carnegie Institution of Washington, 1919–1923) by N. L. Britton and J. N. Rose as a truly monumental work. Although in need of revision it still remains the standard work. A recent reprint has made it available once more. Those taxonomists who denigrate Britton and Rose as "splitters" must react very violently to Curt Backeberg to whom splitting was a compulsive habit. Whether or not Backeberg's classification is generally acceptable to botanists does not detract from the debt that all Cacti enthusiasts owe him for his 6 volume work, *Die Cactaceae* (Fischer, 1958–1962). Its 2,550 photographs alone make it a useful reference work. At one time an English translation was mooted but this seems to have been abandoned. A distillation of this mammoth work was made by Backeberg in *Das Kakteen Lexikon* (Fischer, 1966) which in the more manageable proportions of an octavo volume provides succinct descriptions of all known species of Cacti. What promises to be an exhaustive monograph of the group, if ever it is completed, has been appearing in parts since 1956. This is H. Krainz's *Die Kakteen* with F. Buxbaum providing the morphological notes. Plain or coloured photographs illustrate all the species described.

Trees and shrubs

It is impossible to conceive of our gardens without trees and shrubs as decorative features, focal points or windbreaks. *Trees and shrubs, hardy in the British Isles* by W. J. Bean, former Curator of Kew Gardens, is the most authorititative work on the subject. First published in 1914 in 2 volumes, it is now progressing through its 8th edition which will be completed in 4 volumes. *Trees and shrubs* is deservedly a favourite not only for its masterly presentation of reliable information, taxonomic and cultural, but also for Bean's very personal comments which, fortunately, are still preserved in this latest edition. Its one serious defect is the lack of keys which add so much to the usefulness of A. Rehder's *Manual of cultivated trees and shrubs hardy in North America exclusive of the sub-tropical and warmer temperate regions* (ed. 2, Macmillan Co., 1940). Rehder concisely described over 2,500 species, plus many varieties and hybrids. In fulfilment of his intention to follow this book with "a bibliographical supplement supplying for all the names mentioned in the Manual, exact citations of their sources, and for trinomials the category considered to be botanically correct", he began in his 77th year the *Bibliography of cultivated trees and shrubs hardy in the cooler temperate regions of the northern hemisphere*. It was published by the Arnold Arboretum in 1949. For those who read Dutch, B. K. Boom's *Nederlandse Dendrologie* (ed. 5, Veenman & Zonen, 1965) supplements some of the data given by Rehder. It is an illustrated guide to the identification of trees and shrubs grown in the Netherlands and consists entirely of keys to species, often in their winter state, utilizing bud and shoot characters. A worthy successor to the great German dendrologies by Koch, Schneider, etc. is Gerd Krüssmann's *Handbuch der Laubgehölze* (2 vols., Parey 1960, 1962). It has good accounts of deciduous and evergreen woody plants, excluding gymnosperms, and many line drawings, but lacks keys. The amateur who does not wish to consult such technical tomes may find *The identification of trees and shrubs* (ed. 2, Dent, 1948) by F. K. Makins all he needs to recognize most of the trees he is likely to encounter in this country.

The person who had the temerity to recommend some years ago a "kill a Rhododendron a day" campaign was soon silenced by the indignation of numerous lovers of that spectacular genus. In 1930 the Rhododendron Society brought out the *Species of Rhododendron* which went into a second edition in 1947. It has single-page descriptions of each species grouped under series, and line-drawings of characteristic features. Keys to the species are provided but it is regrettable that there are not also keys to the series. The diminutive size of *The Rhododendron Handbook* (Part 1 1967, Part 2 1969) published by the Royal Horticultural Society belies the extent and depth of information it contains. The cultivated species are graded by an ingenious shorthand notation for qualities of hardiness, size, etc. Also included are the collectors' numbers for Farrer, Forrest, Ludlow, etc. Its painstaking investigation of the parentage of hybrids would impress any genealogist.

Conifers also have their own band of fervent followers. Their bible is W. Dallimore and A. B. Jackson's *A Handbook of Coniferae and Ginkgoaceae* (Arnold), first published in 1923 and revised by S. G. Harrison in a fourth edition in 1966. The genera are now arranged in one alphabetical sequence instead of in families as in previous editions. Having good keys it is likely to remain the basic work for English readers. H. J. Welch, a nurseryman specializing in conifers, has written about *Dwarf Conifers* (ed. 2, Faber, 1968). Despite the author's playful digs at professional botanists he, nevertheless, follows the International Code of Botanical Nomenclature, and boldly attempts to unravel the complex nomenclature of dwarf conifers. Another nurseryman, this time a Dutchman, P. den Ouden, collaborated with the horticultural taxonomist, B. K. Boom, to write a *Manual of cultivated Conifers hardy in the cold-and-warm-temperature-zone* (Nijhoff, 1965). In an alphabetical list they describe nearly 2,500 species, varieties and cultivars. The German conifer authority is unquestionably Gerd Krüssmann. Good descriptions of species, varieties and garden forms are to be found in his *Die Nadelgehölze* (ed. 2, Parey, 1961) with keys which, unfortunately, are absent from his *Handbuch der Nadelgehölze* (Parey, 1972).

Ferns

Reginald Kaye's *Hardy Ferns* (Faber, 1968) which describes suitable species and their cultivation is an indication of a recent revival of interest in ferns among gardeners. If this interest is maintained we will doubtless see a demand for the Victorian classics on pteridology by W. J. Hooker, E. J. Lowe and others.

Publications by societies

In passing I have already referred to the work of some of the smaller horticultural societies but time will not permit me to do more than recommend their publications to your attention. The Alpine Garden Society whose *Bulletin* first appeared in 1930 have now embarked on a series of modestly priced booklets on individual genera presenting a nice balance of horticultural and botanical information. Then there are the periodicals and annuals of such societies as the African Succulent Plant Society, the British Bromeliad Society, the Fuchsia Society, the Heather Society, the Iris Society, the Mammillaria Society, the Rose Society, the Sempervivum Society, etc.

All through this lecture I have assumed that you would not be denied the use of a good horticultural library. Those of you who are members of the Royal Horticultural Society are privileged to borrow books from its Library, which is the most comprehensive of its kind in this country. Its catalogue, although published in 1927, is indispensable for older literature. The catalogue of its counterpart in the U.S.A. was published in 3 volumes in 1962 as the *Dictionary Catalog of the Library of the Massachusetts Horticultural Society* (Hall). These two catalogues capture between them most of the significant contributions to horticultural literature. The dual role, however, of the *Index Londinensis* should not be overlooked, that of providing references not only to illustrations but also to articles and plant descriptions.

The next best thing to growing and observing plants is reading about them, especially on dark winter evenings. May I end by repeating E. A. Bowles' valediction in a similar lecture given to a Royal Horticultural Society audience some 35 years ago and "wish you all good hunting in the book lists and well-filled shelves".

DISCUSSION

LORD TALBOT DE MALAHIDE asked about two new periodicals on the subject of co-ordinated bibliography. He said that those who liked to call their plants by the latest and correct names were very much hampered, for there seemed to be no publication to which they could turn, which gave the latest names. Was there any hope that either of these new periodicals might publish a list of all new name changes.

Mr DESMOND replied that the two periodicals referred to: the *Kew Record of Taxonomic Literature* and Asher's *Guide to Botanical Periodicals* would not list new name changes. Asher's *Guide* would be simply a listing of the contents pages of all the periodicals that they proposed to scan, about 5,000 he thought. No attempt would be made to analyse the contents of the articles in the periodicals. The *Kew Record* would list, in systematic order, all the literature for the period under survey. It would be an annual publication and in it would be included all new names in the literature that had been validly published, but not all name changes.

There was no publication which gave the latest correct names nor could he imagine that any was envisaged. The compilation would be a colossal task to undertake. If one wanted to establish whether a name had been validly published, one would first go the the *Index Kewensis,* but this was something which was only available in libraries and there was nothing which was handy, portable and available for the home.

Dr H. HEINE said there were the useful Hand-lists of Conifers and Orchids put out by Kew. Unfortunately there were only these two but if there had been more they would have been exactly what Lord Talbot was looking for. If the Kew Seed-list, which was well prepared, with up-to-date nomenclature, was accessible to all those interested in horticulture, it would make a very great contribution towards the stability of names for garden plants. But usually the Seed-list was accessible only to similar institutions to Kew, and contained a limited number of plants.

Mr R. LEWIS felt there was one gap in the topics Mr Desmond had dealt with, and that was catalogues, whether old or new, from nurseries or from botanic gardens. Could he describe the availability of those catalogues which were known not to be available to all, yet contained the kind of descriptions of plants that gardeners wanted? One could turn to the enormous lists made by botanists, but one came away clouded by descriptions and technicalities. Very often what one would like would be lists of plants, extensive lists certainly, but with details of the differences between plants that may be immediately noticeable in the garden, and so give gardeners an idea of what one would plant to obtain different forms or different flowering periods, and other practical observations of this sort.

Mr DESMOND replied that nurserymen's catalogues were too often treated as ephemeral material. There was a tendency to discard and throw them away. It was only in libraries like that at Kew and the Royal Horticultural Society, and perhaps in a slight way the Natural History Museum, where they were kept. Although there was a small collection of nurserymen's catalogues at Kew, the best collection was undoubtedly that of the R.H.S., but that too was very fragmentary and went back to about the beginning of the 19th century. John Harvey had recently published a book, which you may have seen, entitled *Early Gardening Catalogues* (Phillimore Press). Harvey had spent three years or so trying to locate catalogues up and down the country, either in private possession or in university libraries, and had found, as he had anticipated at the beginning, that good collections just did not exist. There was one old catalogue here and another there. Mr Desmond said he had been in Holland recently, and at the Dutch Bulb Growers' Association in Haarlem had found that they had one of the best collections of British nurserymen's catalogues that he had come across, apart from that at the R.H.S. It had been suggested to the International Documentation Centre that they should produce a microfiche of all the available nurserymen's catalogues, in the same way as they had recently published one of seed-lists, which contain vital information to botanists. They were now considering whether this was a viable proposition, and it may be that in a year or two these catalogues, such as they were, would be available to the general public in more libraries than just the few he had mentioned.

HEBE

P. S. GREEN

Royal Botanic Gardens, Kew

Introduction

To many of you, I suspect, the plants I have been asked to speak about are known as shrubby Veronicas. However, the name *Hebe* is gaining general acceptance and I believe you will agree that it is one of the easiest of botanical names to remember.

Two different Hebes are naturalized in Britain, both escapes from cultivation, for in our gardens we grow over 40 species and a host of hybrids. In a quarter of an hour I cannot attempt a proper account of *Hebe*, nor am I competant to even try, for although it is a group I would enjoy studying, I have never been able to more than dabble in it from time to time. As he had turned to me for help over the nomenclature of one of the two plants naturalized in Britain, the President of the Botanical Society twisted my arm and persuaded me, against my better judgement, to try and fill the gap in the programme left by the untimely death of Dr Donald Young who was to have talked on *Oxalis*. I fear this will be a poor substitute.

Although the genus *Hebe*, which is now thought to contain about 100 species, was first described as long ago as 1789 it was not until the 1920s that it was really taken up. First by Pennell, when revising the New World species of *Veronica,* and then shortly afterwards by the two New Zealand botanists Cockayne and Allan who brought together an account of the species of that country, which is the main centre for the genus, and published a revision in 1926. Although the species had previously been treated as constituting a section of *Veronica* their separation at generic level is based upon good characters of the capsule, general habit, geographical distribution and chromosome number.

Horticultural value

Horticulturally speaking, *Hebe* is an important and valuable genus, especially in gardens near the sea. As small evergreen shrubs many make good rock garden subjects, others are noted for their gay white, purple or bluish flowers in summer or early autumn, while others, the "whipcord" Hebes, so-called because ot their closely appressed and reduced leaves, resembling some species of *Cupressus,* are grown for their unusual and curious habit. The species hybridize very readily and because of the taxonomic problems this creates I hesitate to say whether this is something in their favour or not. From the time they were first introduced into cultivation, now over a century ago, they have been crossed by gardeners or have even hybridized spontaneously. Important from the gardening point of view too is the fact that these hybrids may be readily propagated by means of cuttings. Even in the wild, in New Zealand, they hybridize freely and frequently, presumably as a result of the removal of ecological barriers between species consequent to the clearing of forests, or other activities of the pioneer farmers and settlers. A great number of named hybrids have been produced and many of these are still in cultivation. As you can imagine, their identification can be very difficult and, although some of the confusion has been cleared, there are still uncertainties and problems of identity amongst the garden hybrids and cultivars.

The classification of Hebe

A great advance in the classification of *Hebe* was made with the publication of volume 1 of Allan's *Flora of New Zealand* in 1961. In this Dr Lucy Moore has presented an outstandingly useful account of the genus in New Zealand, which after all is its main centre. Based on extensive field work, cultivation experiments and critical herbarium studies, her revision recognizes 79 species and a number of varieties. As well as descriptions of individual species, often with critical comments, Dr Moore has provided a workable Key, extensive notes on hybridization in the wild, and a section headed "Horticultural forms."

Two naturalized in Britain

Although there are numerous Hebes in our gardens, only two of them have become naturalized in Britain. *H. salicifolia,* a native of both New Zealand and Chile, has undoubtedly escaped from cultivation, for it is a well-known and valuable garden plant.

The other naturalized *Hebe* is a hybrid, not a species, yet in spite of this is quite fertile, sets seed and spreads itself by this means. As with several Hebes it is resistant to sea-spray and has been widely used as a hedge plant and low wind-break in seaside gardens. In Dandy's *British Plant List* and in Clapham, Tutin & Warburg's *Flora of the British Isles* this hybrid is listed as *H. × lewisii,* but a recent investigation has shown that this name has been misapplied and really belongs to a hybrid of different parentage. The parents of the naturalized plant are *H. elliptica* and *H. speciosa* and the cross was first made in the garden of Mr Isaac Anderson-Henry in Edinburgh just over a century ago. He proposed the name *Veronica × lobelioides* for the resultant plant, but although it appears in the literature this name was never properly published with a description, and relatively early on, but quite incorrectly, the epithet *lewisii* came into use for this plant instead. Many years later, Miss Alice Eastwood, working in California, found the need for a name for this hybrid now growing, and in fact naturalized as well, in and near San Francisco. She failed to find a valid name for the hybrid and called it *Veronica × franciscana.* A few years later, Mr Souster, then of Kew, transferred this epithet to *Hebe* and there seems no doubt that this is the correct name for the plant which is now naturalized in the West country, in the Isles of Scilly, Cornwall and Devon, as well as in the Channel Isles.

A worth-while genus to study

These two Hebes, although naturalized in a number of places, are probably insufficiently hardy to compete with native plants, except under favourable and local climatic conditions. Many of the species are only marginally hardy in much of Britain but a lot are quite at home in our gardens, and if anyone with garden space is looking for an interesting and challenging genus to collect and study I thoroughly recommend *Hebe*. Many of the cultivated varieties and hybrids badly need critical and comparative study and here is the opening for someone of a fascinating and worth while hobby.

ARUMS—LORDS-AND-LADIES

C. T. Prime

Introduction

I have been asked to "explain the delights of wild Arums to gardeners and of garden Arums to botanists". In attempting this in ten minutes I may perhaps remind you that the *Araceae* are a large and mainly tropical family of over 2,000 species. There are three species native to Britain and one or two that have been introduced and have just about gained a foothold. The family includes some house and greenhouse plants, like *Anthurium, Monstera, Spathiphyllum* and *Philodendron;* some garden plants like *Dracunculus* and the Arisaemas; one plant of some horticultural importance, the Arum Lily (*Zantedeschia aethiopica*), which is widely used for decoration at Easter time; and some aquatics, the Cryptocorynes, which find a place in tropical fish tanks. Time only allows me to talk about one or two species of *Arum* proper.

Arums for the gardener

Firstly, for gardeners I can recommend the white-veined forms of *Arum italicum* (Plate VI). This plant, which Miller says, "grows naturally in Italy, Spain and Portugal from whence I have received the Seeds", was first described by him in 1759. He further says, "The Leaves of this Sort rise a foot and a half high, are very large, running out to a Point; these are finely veined with white, interspersed with black Spots, which together with the fine shining green of their Surface, make a pretty Variety....". *Arum italicum* is a native of the south and west of England, but usually the plants lack the conspicuous white venation and are collectively known as *A. italicum* subsp. *neglectum*. Botanically, however, the white-veined forms are known as *A. italicum* subsp. *italicum*. The Royal Horticultural Society's *Dictionary of Gardening* lists two horticultural varieties, var. *pictum* (which means painted) and var. *marmoratum* (which means marbled). I am not very clear as to the difference between these, but Engler (1920) did describe a variety *marmoratum* as having both black spots and white veins, so presumably var. *pictum* has the white veins only.

The plant has two advantages, one that the foliage is winter green, for the leaves appear in October when most other plants are dying down, and secondly, it is easy to grow—almost too easy, for in favourable soils it may become a weed. The flower spike does not attract everyone, but the berries are quite lovely in autumn. The plant is quite easy to obtain.

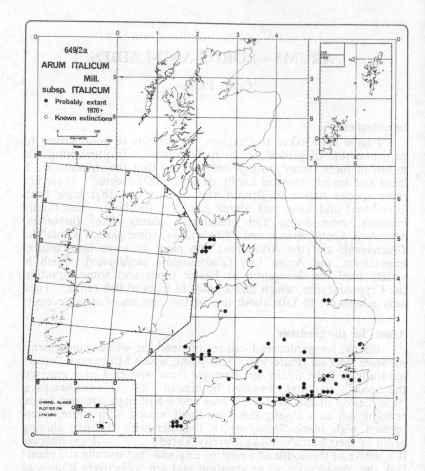

FIGURE 1
The distribution in the British Isles of *Arum italicum* subsp. *italicum*.

The common native species of Great Britain is *Arum maculatum,* so named because it often has spotted leaves. Spotting is much rarer in the leaves of *A. italicum*, just mentioned, and is perhaps evidence of hybridity, for although *A. maculatum* occasionally hybridizes in England with *A. italicum* this does not happen all that frequently, because the two species do not often occur together and their flowering periods are different. However, Plate VII shows a cross between a very spotted *A. maculatum* and *A. italicum*. I hope you agree that it is a very striking plant and worthy of a place in the garden, if you can get hold of it.

Secondly, I would recommend to the gardener one other species, *Arum pictum* proper, a native of the Balearic Isles, Sardinia and Corsica. This is unlike all the other species of *Arum* in being autumn flowering; it is also winter green. So although you may not wish to have a spring flowering species in your garden when there so much that is glorious to grow, you may not despise an Aroid in autumn when you have little else like it in flower. The flower spike is a deep maroon and the foliage appears about the same time; it is different from *A. italicum* in being only slightly marbled, but it is as attractive in its own way, if not quite so spectacular. It is, in my experience, not so easy to grow, but if you have a well drained soil against a south facing wall, it is worth a try. The Royal Horticultural Society's *Dictionary* describes it as hardy, and I have grown it outside for some years, though not to flowering. It is not too easy a plant to obtain.

Another quite delightful species is the variety of *Arum creticum* with yellow spathes and long appendices to the spadix (Plate VIII); I think it is best treated as a pot plant for the cool greenhouse, for I must confess I find it a little disappointing in that it will produce many small corms and fail to flower.

Arums for the field botanist

So much for suggestions for the gardener; now for the field botanist. The subspecies of *A. italicum* with white veins has escaped into the countryside, in most cases it seems, by being thrown away with garden waste rather than by spreading from seed. It is therefore found in the neighbourhood of buildings, and the map (Fig. 1) shows the occurrence of the cultivated subspecies in the wild, so far as is known at the present. In some places it has only persisted for a short time, but in others it has been of long standing.

We may perhaps trace the history of the spread of this plant into the countryside. The plant had been recognized before Miller's time; it was probably the *Arum venis albis* of Bauhin, but Linnaeus did not regard it as specifically distinct and included it within the common native species, *A. maculatum*. It is included in the list of plants cultivated in the Edinburgh Botanic Garden in 1683 under the name *Arum venis albis B.P. magnum rotundiore folio* Park. (Round leaved Wake-Robin), though the identification is suspect, for Parkinson (1640) says of *Arum magnum rotundiore folio*, "This kinde hath somewhat larger leaves than either of the former and more round pointed, both at the end and at the bottom next the stalke, having some white veins appearing in the leaves, and abiding greene longer in the Sommer, even almost untill Automne, the hose or huske with the pestill or clapper are both of a pale whitish yellow colour, in which things this differeth from the other and in nothing else". One would say that the leaves of *A. italicum* are more acute than those of *A. maculatum;* moreover, he goes on to say that these plants "shoote forth leaves in the Spring and continue but until the

middle of Sommer" which does not fit with the winter green habit
of *A. italicum.* The *Hortus Kewensis* of 1769 lists eleven species
but not *A. italicum.*

It seems most likely that *A. italicum* subsp. *italicum* found favour
as a garden plant on the recommendation of P. Miller. The
distribution of the escapes is of some interest, being mainly southern
and western. It resembles the distribution of the native subsp.
neglectum (Fig. 2), which there is reason to believe may be climati-
cally determined, and this raises the question as to whether the same
applies to subsp. *italicum.* In an effort to shed some light on this,
I tried to gain some information on how the plant was spread about
the country. Looking at some twenty old seedsmen's catalogues
and lists in the Lindley Library dating from the early 19th Century,

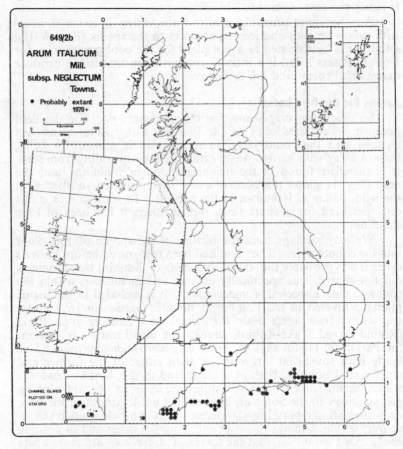

FIGURE 2
The distribution in the British Isles of *Arum italicum* subsp. *neglectum.*

A. italicum occurs in most of them, but it does not occur in three of the lists from Edinburgh and Newcastle although they include some other *Arum* spp. On this slender evidence it seems that the plant was not sold so readily in the north as in the south of Britain. The lists also show the plant to have been most popular about the middle of the 19th century and this would explain its present occurrence in the neighbourhood of buildings of that period, rather than those of more recent date. In 1862 Mr John Salter of Versailles nursery, Hammersmith, was awarded a 3rd prize of 15/- by the Royal Horticultural Society for a collection of hardy varieties of variegated plants, including *A. italicum*. The plant is also mentioned in a few other connexions in the Society's *Journal* about that time, but after that it suffered an eclipse, until recent interest in flower arranging has brought it once more to the fore.

So we are still left with the problem of explaining the plant's distribution. Is it due to the distribution of plant salesmen in the 19th century or to some more natural factors? That is my reason for asking field botanists and gardeners to keep an eye open for this plant as an escape, particularly in the winter when it is easy to see, in order to make the map complete and accurate. If the plant has not been distributed in the north, the map does not tell us very much; if it has been and now does not succeed then the distribution must be caused by some other factors. Should the maps of subspp. *italicum* and *neglectum* end up being similar, this would suggest that subsp. *neglectum* has reached its natural limit and is not likely to spread further. Thus the escape of a garden plant, can in a sort of way serve as an experiment to elucidate the factors causing the distribution of a native species. There are some flaws in this argument as presented which there is not time to discuss, but let us have the facts and such conclusions as can be drawn from them for what they are worth.

In conclusion I would like to express my thanks to the Biological Records Centre for the two distribution maps.

References

ENGLER, A. (1920). *Das Pflanzenreich* 73 (IV 23F). *Araceae-Aroideae & Araceae-Pistioideae.* Leipzig.
MILLER, P. (1759). *Gardeners Dictionary,* Ed. 7. London.
PARKINSON, J. (1640). *Theatrum botanicum.*

DANDELIONS

A. J. RICHARDS

University of Newcastle

Introduction

Earlier in the meeting Mr Meikle has expressed astonishment that natural selection has not left "a murkier trail of. . . . ill-defined and misbegotten species" and other groups. Most botanists would agree that this could scarcely apply to the species of *Taraxacum,* members of a very large, successful, widespread and familiar genus in the family *Compositae,* which has justly earned a reputation for taxonomic complexity.

It is possible to split the history of the study of *Taraxacum* into an "Anglo-American School" and others. Until 1970, only one species of *Taraxacum* had ever been legitimately described by a British author (by Lyons in 1763 as *Leontodon palustre,* incidentally the earliest available basionym in the genus), and it is to the Swedes and Finns that we owe our modern knowledge of the genus. Indeed, the only critical work on the not inconsiderable but still largely unknown native *Taraxacum* flora of the Americas is by a Swede, Gustav Haglund, despite the work of Nelson, Rydberg, Fernald and others.

Handel-Mazzetti (1907) published the first monograph of the genus, but this considers only 56 species of a very broad nature, many of them equivalent to modern sections. These "macro-species" each cover a wide range of morphological, genetical and ecological entities and are of little help to the serious student of the genus. Nevertheless, up to now they have been almost exclusively employed in Britain and America.

Apomictic species

By 1907, however, it had already been shown that many Dandelions do not produce seeds through a sexual process, but are apomictic. That is, during the development of the ovules there is no reductional meiosis, nor later any fertilization or sexual fusion. As a result, the progeny are, genetically speaking, identical to the parents and a large number of invariable clones are propagated by seed— units which are not easily classified into groups. A number of Scandinavian and Dutch authors have set out to describe each of these clones as a species (a microspecies or an agamospecies, that is, a species lacking sexual fertilization), classifying them where possible into sections (the most useful taxon for grouping aggregates of agamospecies). These microspecies now number nearly 2,000 (132 in the British Isles) and despite their multiplicity have proved a much more satisfactory category than the macrospecies. For a useful taxonomy in groups like *Taraxacum* it is necessary to treat complexity in a complex manner, and the microspecies are perhaps not over-many for a genus native in five continents and occurring in a remarkably wide range of habitat types.

Taxonomically there can be no doubt that *Taraxacum* is a critical genus, but this can be attributed to a lack of monographs and floristic accounts, to the very considerable plasticity inherent in many species and to the complex breeding system, rather than to the sheer numbers of species.

After Murbeck, in 1904, showed that *Taraxaca* were seed apomicts, a considerable body of cytological and genetical work was undertaken in the genus, principally by Gustafsson. This revealed many cytological races, from diploid to hexaploid, with many aneuploids. The diploids, which mostly occur in warmer areas in the Mediterranean region and in Asia, proved to be entirely sexual (and later work has provided the additional information that they are almost all self-incompatible: the reason why, in most circumstances, they are readily detected by poor seed-set). All polyploids were thought to be apomictic, with uneven chromosome sets and very disturbed meioses. In fact, many even lack pollen completely. Their apomictic behaviour is readily demonstrated by anyone using a razor to remove the upper two-thirds of a bud. The seeds still develop; indeed embryo development normally proceeds before the flower opens, thus precluding the possibility of fertilization. Recent work (Sørensen, 1958; Richards, 1970a) has shown that a few triploids which have relatively regular meioses are subsexual and most of these produce some sexual and some apomictic ovules in the same flower-head. Both sexual and sub-sexual Dandelions are uncommon in western Europe. When they do occur, hybridization with apomicts bearing pollen, as well as with other sexuals, is commonplace, and in this way new apomictic genotypes arise (Richards, 1970b; and unpublished), yet these rarely seem to be sufficiently successful to warrant recognition by a specific name.

Plasticity

Taraxacum has long been renowned for the plasticity of its individuals: that is to say, for the capability of one genotype to exhibit different phenotypes with age, or as a result of environmental modification. It appears that the ontogenetic sequence of the leaf forms which are produced during the life of a plant, and which are to some extent recapitulated every season, mirrors the ortho-genetic series of leaf morphologies which can be observed in the evolutionary development from primitive to derived species within the genus. The most primitive types usually have simple leaves, or leaves with short, simple lobes, which recall those of *Leontodon, Scorzonera* or *Crepis* more than the familiar *Taraxaca*. With more derived species we encounter increasingly complex leaf-forms, exhibiting greater dissection, and more varied tooth, interlobe and lobe characters. Juvenile leaves in these derived types are simple, and usually unlobed. As the plant ages, more complex leaves are developed, yet it can be observed that those produced on an adult plant during the first flush of flowering, which are formed during the previous autumn and remain in an unexpanded state during the winter, are relatively simple. After flowering, more complex leaves are produced, and these are more liable to be influenced by the environment. They also show a greater resemblance between species than do the spring leaves and, as a result, these later forms (known as "*status aestivalis*") are not usually of taxonomic use. This increasing complexity in leaf morphology during the life of a plant, or during a season, can be blocked or hastened in extreme environments. Thus, shaded forms remain in a juvenile, entire state, while those in dry or trodden localities spend most of the season in a highly dissected condition. Neither are readily identifiable, and taxonomists thus suffer the handicap of requiring well-preserved specimens from typical, well-grown material in spring (specimens in which colour has been preserved by careful drying). Nevertheless, even in suitable environments, the amount of plasticity exhibited is considerable, and the "art" of identifying plants correctly depends largely on the possession of sufficient familiarity with the species to be aware of its plastic capabilities. As a result, when identifying collections, much attention has to be given to the rather difficult characters of the achene, involucre and capitulum, each of which is relatively stable. There is no substitute for comparison with good material in a herbarium, and one of my objectives is to ensure that well-collected, correctly identified specimens are available at a number of centres in this country (Richards, 1972a). However, I have little influence over the quality of material generally collected, and I would urge British collectors to examine typical Scandinavian specimens (at Oxford, Kew or the British Museum) before adopting standards they think are adequate!

There can be little doubt that the exceptional capacity for plasticity found in Dandelions is associated with its apomictic reproduction. It seems likely (Richards, 1973) that apomixis arose rather early in the history of the genus, certainly before the start of the Ice-Age, and that these apomicts, no longer capable of genetic variation, survived best when possessed of considerable non-genetic plasticity to exploit a greater number of environmental opportunities. It is clear from cytological and genetical results that most, if not all, apomictic Dandelions are of hybrid origin (Richards, 1972b, 1973). It may well be that most of the familiar species of western Europe, which have become so firmly established as weeds in many parts of the world, originated as a result of hybridization between ancient, arctic, apomictic species and hybridizing sexuals. A vast hybrid swarm would have thus been "fixed" in perpetuity by apomixis. We view today the discontinuities in this "frozen" swarm which have survived the ravages of natural selection, and quite properly call them species. The more conventional sexual species are the result of natural selection working on genetic variation, in a similar manner. As the number of apomictic genotypes became fewer, so the requirement for plasticity may have increased, and a major component in the fitness exhibited by the surviving types may have been their capacity for such phenotypic variation.

Vigorous weeds

A characteristic of hybrids is their vigour: they grow more quickly, to a larger size, and show a greater capacity for vegetative reproduction than their parents. In many instances this latter characteristic may be their salvation, for they are frequently sterile. Yet in their vigour, and in their unique combination of quite different genetic complements, potentially they are often very successful plants, hampered only by their sterility. This handicap may be overcome by apomixis, so aptly termed by Darlington "an escape from sterility". In fixing the vigour and genetic innovation inherent in hybrids, apomixis may change the nature of the plant's potential. Due to the abolition of the need for sexual reproduction, with its accompanying hazards, it becomes better suited to survive in extreme environments. However, it will have lost the highly evolved and complex genetic mechanisms which control "niche specificity", which allow a species to compete in a closed and highly structured community. It therefore becomes more suited to open ground, where its vigour, ready seed-set (and dispersal in the case of Dandelions), and lack of ecological specificity suit it admirably. Thus, many *Taraxacum* species which today are rather widespread in arable and other open habitats may originally have been of very local occurrence, limited to the fewer open habitats provided by nature. Many such species show an anomalous distribution attributable to the accidental introduction of an originally localized species to suitable artificial habitats in quite different geographical areas (Richards, 1972a).

It is scarcely surprising that many Dandelions are successful weeds. They are vigorous, and very difficult to kill. They are fast-growing, and will flower in less than a year from germination, often much sooner, yet are long-lived perennials. They produce abundant seeds, all fertile, all equally fitted to their open environment, seeds which are most effectively dispersed, and which germinate quickly and readily at any time of year. Dandelions set seed early in the year, before competition for open ground becomes too keen, while, in addition, they are capable of vegetative reproduction from very small segments of their root, which readily produce adventitious buds. And they are sufficiently plastic to accommodate to a wide range of environments. These are the very qualities of a "super-weed".

Useful Dandelions

Yet, *Taraxacum* is at the same time a most useful genus. It is medicinal, forming a most effective diuretic. It makes tasty drinks, and savoury salads. And, for a brief period of history, one species may have played a significant part in the winning of a World War. The Russians were isolated from tropical sources of rubber during much of the 1939–45 war, and for this period over 70% of their rubber needs were met by an aneuploid (2n=25), subsexual hybrid between the sexual, Asian *T. kok-sagyz* Rodin and a species in the section *Vulgaria* Dahlst. These plants were cultivated in very large quantities in the Crimea and Georgia, and to a lesser extent in Byelorussia and the Ukraine. Special machines were built to harvest the "Rubber Dandelion" and factories put into operation to extract the rubber. It is claimed that over 90% of the latex of these plants was rubber (Krotkov, 1945). Later, the advent of cheap sources of synthetic rubber, this rather laborious, but at the time essential, means of obtaining rubber fell into disuse.

Dandelions as gardens plants

Taraxacum has yet to make an impact on the horticultural world of the West. In Japan, where the native species are less troublesome as weeds, no such inhibitions exist, and the cultivation of decorative species has led to the formation of a Dandelion Appreciation Society. The genus is varied, not only in leaf shape and texture, but also in flower colour. Some of the colours encountered are listed:

Purple	*T. porphyranthum* Boiss.
Rose	*T. roseum* Hand.-Mazz. and *T. pseudoroseum* Schischk.
Lilac	*T. liliacinum* Krassn. ex Schischk.
White	*T. arcticum* (Trautv.) Dahlst., *T. leucanthum* Ledeb., *T. dealbatum* Hand.-Mazz. and *T. albidum* (Makino) Dahlst.
Ochre	*T. cucullatum* Dahlst.
Copper	*T. aurantiacum* Dahlst.
Orange	*T. croceum* Dahlst. and relatives.

In leaf characters and habit, also, a fascinating range of forms is to be encountered, some of great potential to the horticulturist. There are species with leaves that are farinose (*T. serotinum* Poiret), somewhat woolly (*T. montanum* (Meyer) DC.), entire and fleshy (*T. obovatum* DC.) or grass-like and linear (including the British native *T. palustre* (Lyons) Symons). Nor need many of these plants be avoided because of invasive habits. Although the "weed" species, few of which have any aesthetic merit, cause problems for the gardener by virtue of their vigour and abundant seed-set promoted by apomixis, the majority of those which are "desirable" are sexual diploids, and are self-incompatible. As a result they set little seed, and they lack the vigour of the polyploid hybrids. Indeed, many are difficult to grow, and in order to survive for more than a short period in an English garden some require greater skill than I can muster.

Even in cold greenhouses many are tricky plants, liable to attack by aphids, eelworm and root-bugs; and readily rotted or damped-off by overwatering. In addition, some are difficult to flower, this being particularly true of species from dry habitats in central Asia. The reasons for these vicissitudes are not hard to discern. Many Asian plants are adapted to conditions where there is continuous water at the roots during the growing season in early spring, followed by total drought through the summer and autumn. Others, from alpine habitats, are accustomed to plentiful supplies of water and sun in the summer, and cold, dry conditions during the winter. It is well known among gardeners that plants such as these are amongst the most difficult to cultivate in the equable, damp climate of the British Isles, and this is certainly true of many species of *Taraxacum*. Knowing the stimulus and pleasure many gardeners receive from trying to cultivate difficult plants, I hope this may act as a recommendation rather than a discouragement for attempts to grow a number of species in this neglected and fascinating genus.

REFERENCES

HANDEL-MAZZETTI, H. VON. (1907). *Monographie der Gattung Taraxacum.* Leipzig & Wien.
KROTKOV, G. (1945). A review of literature on *Tarazacum kok-saghyz* Rod. *Bot. Rev.* **11**: 417–461.
RICHARDS, A. J. (1970a). Eutriploid facultative agamospermy in *Taraxacum. New Phytol.* **69**: 761–774.
——— (1970b). Observations on *Taraxacum* sect. *Erythrosperma* in Slovakia. *Acta Fac. Rerum. Nat. Univ. Comen.-Botanica* **18**: 81–120.
——— (1972a). The *Taraxacum* Flora of the British Isles. *Watsonia,* **9**: supplement: 1–141.
——— (1972b). The karyology of some *Taraxacum* species from alpine regions of Europe. *Bot. J. Linn. Soc.* **65**: 47–59.
——— (1973). The origin of *Taraxacum* agamospecies. *Bot. J. Linn. Soc.* **66**: (in press).
SØRENSEN, TH. (1958). Sexual chromosome aberrants in apomictic *Taraxaca. Bot. Tidsskr.* **54**: 1–22.

DISCUSSION

Mr I. R. BONNER asked if one had sexual British species, was it possible to cross these with the pollen of others which were apomictic and so start a new clone or new microspecies?

Dr RICHARDS replied that it was perfectly possible and did happen. In Britain sexual species were not very widespread, in fact the two wholly sexual species were very local indeed and quite recently introduced. They had actually been found only once. The native sexual species were not always sexual, only sometimes. In localities where they were sexual one found hybrids with the apomicts, which had acted as pollen parents. Many of these hybrids were themselves apomictic, so one could say that new species had arisen de novo. However, in all the localities where he had seen this happen in this country and abroad, except in one locality in Czechoslovakia, it appeared that the hybrids were actually unfitted for existence and usually did not survive in the wild for long. There was no indication that new species were being formed in large quantities. However, in the one locality in Czechoslovakia there was very good evidence that new species had arisen this way quite recently.

Mr E. MILNE-REDHEAD asked whether there was any reason why the asexual British species should be very persistent while, in contrast, the sexual Japanese ones were difficult to cultivate?

Dr RICHARDS replied that the answer probably lay in the fact that most of the sexual Asian species (which were largely from Central Asia, although some were Japanese) were alpines in the true Alpine Garden Society sense. They were difficult rock garden plants, with a propensity to dampen-off in winter around the crown, and they were particularly susceptible to fungal attack. He thought that one would probably be able to cultivate them if they were treated as plants for the alpine house; indeed in his experimental greenhouse he had grown them placed in a well-drained position with rock on the top and so on; but they do dampen-off.

Mr MILNE-REDHEAD said he thought it would be very useful if they could be crossed with the common weedy Dandelion of our gardens so that these damped-off in winter!

Dr H. HEINE asked why, when the genus *Taraxacum* was cosmopolitan, there were no species listed as aliens in Dr Richards's interesting account of British Dandelions [*Watsonia* 9, Supplement: 1–141 (1972)]. Surely they had been introduced from other parts of the world. Was this lack of a list of non-native species due to a still imperfect knowledge of these apomicts in other countries, or was it still too early to recognize whether they were aliens or not?

Dr RICHARDS replied that there were 132 species of Dandelion currently recognized in the British Isles. Of these he thought that only about 85 were native. In his account he had said there were a few species of uncertain status, not known to be native or introduced. In many cases one could guess: if one had a single record from a rubbish tip or a dockside then the plant was almost certain to be an introduction. In yet other cases, where plants were apparently not very permanent in their locality, and showed slightly scattered distributions, one could have a very good guess that they were probably introduced. He estimated that something like one quarter of the species recorded in Britain were not native to the British Isles. Dr Heine was quite right in suggesting that there were large areas of the world in which *Taraxacum* occurred but was not native. In parts of the tropics, the *Vulgaria* were really very common, he had been told, in particular the weedy areas of South America and quite large portions of Africa and SE. Asia. In the high altitude, temperate areas of the tropics and in the Southern Hemisphere, there were some native Dandelions but many European species were adventive throughout the world. In temperate North America, where Dandelions were only native in the Rockies and one got Arctic species which came down the mountain chain, there were quite a large number of European species adventive in other parts some distance from these regions.

SUMMING-UP

David McClintock

President of the B.S.B.I.

I believe we all have the strong feeling, not only that we have had a very good gathering, but a very distinguished gathering. We have had people with great experience speaking and in the audience, and my regret has been that since most of the speakers talked too well, there was not time enough for discussion, when even more good points would have been made. But I do not think we could have crammed in more goodness. Frances Perry last night remarked on the number of younger people in the audience, which I too was delighted to see. We were very sorry about all those—150 or more— we had to turn away. This was a deliberate decision. We thought this lecture room full—230 people—was a more manageable number, and elsewhere in these halls slides would have been impossible and catering and control more difficult. When we plan the next conference, we shall have to decide whether it should also be restricted to this size or whether we can have a large one, or what. There is going to be a report published on this one, in a hard-back book, so you will all get the chance of reading the papers. I found sometimes I could not take in all the copious facts and details, so this will be valuable for refreshment and for reference. It will be announced to both our Societies when it will be ready, early next year; and we are very grateful to Mr Peter Green of Kew for agreeing to edit it— and no-one will do it better or more promptly.

My job now is to review two days of almost solid talk from no less than 30 speakers including our eminent Chairmen, all of whom deserve our warm thanks. Each made telling points which I must leave you to refer to in the printed report. But I do feel great gratitude for excellent variety, good humour and a delightful scattering of accents from around the British Isles; the whole thing has been so thoroughly enjoyable and worthwhile.

My chief impression is that we have found more scope and more chance and more reason for close co-operation between gardeners and field botanists than even I thought. The extent to which they had not even entered each others territories was brought home to me strongly when I learnt that some office holders and committee members of the Botanical Society had never even been in these halls before. Now they have!—and will return. We had so many things brought out which members of both societies have in common and in which they can help each other: they are inter-dependent. They have immense goodwill in common. Wherever one goes, one gets remarkable friendliness and co-operation amongst plantsmen of all sorts, whether they are gardeners, or whether they are "wild" people.   I see quite a bit of other naturalists—bird people and bug people and so on, and they are unfailingly friendly to me, but somehow I never feel there is quite the same friendship within their groups as there is among plant people.   It certainly extends across whatever frontiers may be imagined between gardeners and botanists; and any botanists who go to gardens, and I strongly recommend them to do so, should get just as great a welcome as any gardeners who go on our field meetings, and I strongly recommend them to do so too. Indeed we are delighted that well over 100 of you will be on the field meeting tomorrow.

I was talking with Lord Aberconway when we mentioned the wild and naturalized plants of Bodnant.   He suggested we came and looked at them, and I hope the B.S.B.I. will, and visit other such gardens too.   We had a meeting at Wakehurst a year or two ago. noting the wild plants that grew there, as well as seeing the cultivated ones.   But field botanists and gardeners have a lot to gain by just keeping their eyes open, looking at the variants, looking with greater understanding at the weeds that grow, looking for good plants everywhere.   This is something that both should enjoy wherever they go, botanists noticing varieties of Snowdrops, gardeners the different Dandelions, and so on.

One subject we did not go into was the actual and potential importance of public parks and public gardening, where so much can be done for our plants and to foster interest in them—but there was plenty else we could not include in our two-day conference; and in any event some of you may have thought we scampered too fast over what we did include.   So such groups as trees or roses or heathers or orchids or alpine plants must await another meeting; so must photographing and drawing plants.

Another matter where gardeners and botanists have everything in common is conservation: we all want to preserve for posterity our plant heritage, in and out of gardens.   No-one knows when we may not want it for genetic or medicinal reasons, or just for beauty and the diversity it represents.   I am sure field botanists should make much more use of nurseries, so many of which grow native species.   There are much better plants to be got from nurseries than direct from the wild.   Private plantsmen grow a vast range of species which they exchange among themselves, and no doubt will continue to do so; plants which never find their way into catalogues. I was with a panel this summer with Mr Arthur Hellyer, when he remarked that the best way to keep a plant is to give it away.   How right he is.

Another outcome of this conference is that the R.H.S. are nominating an official representative to the Conservation Committee of the B.S.B.I., a move that is warmly welcomed.   There is already considerable representation the other way, for example many members of the R.H.S. Scientific Committee also belong to the B.S.B.I.

One cause of the success of this conference has been its excellent organization. This is in basic measure due to Mr C. D. Brickell and Mr J. E. Elsley of Wisley, and to Mrs M. Briggs of the B.S.B.I. But they have had very good helpers, especially during these two days and for the field meeting tomorrow. I must also thank the exhibitors in the hall for what they put up, and particularly Mr W.K. Aslet for the fantastic hours of labour he and his assistants spent over the Wisley exhibit, and Mr Harold Hillier for his very fine and appropriate show of cultivars of British trees and shrubs. And, once more, I thank our excellent speakers, and you all for coming.

APPENDIX I

EXHIBITS STAGED AT THE CONFERENCE

Nine exhibits were staged in the Royal Horticultural Society's New Hall illustrating some of the themes of the Conference and some of the papers which were presented in the lecture hall.

ALCHEMILLA

Several living plants of a number of Lady's Mantles were exhibited, some of them native in Britain but all from cultivation.

A. alpina L. sens. str.
The true Alpine Lady's Mantle, locally common on British mountains, but not very easily cultivated in lowland rock gardens.

A. conjuncta Bab.
The commonest "Alpine Lady's Mantle" in cultivation, native of Jura and the W. Alps, and wild in Clova, but probably introduced there.

A. faeroensis (Lange) Buser
A remarkable *Alchemilla* belonging to a small group of species intermediate between the '*vulgaris*' and the '*alpina*' groups, and almost certainly of ancient hybrid origin.

A. faeroensis var. *pumila* Rostrup
A very attractive dwarf variant of *A. faeroensis* which retains its habit in cultivation. Not, apparently, difficult to cultivate, and worth a place in specialized 'alpine' collections.

A. glaucescens Wallr.
A neat silvery-hairy species native on limestone in NW. Yorks. (also very locally in Scotland and Ireland). Much more easily cultivated than other native British species, and a very suitable plant for a limestone rock garden.

A. sericata Reichenb. sens. lat.
A rather variable Caucasian species, not common in gardens. Can be told by the sub-appressed, silky indumentum and the leaf-lobes separated by rather deep toothless incisions.

A. mollis (Buser) Rothm.
The commonest *Alchemilla* in gardens, very easily recognized by its shallowly-lobed, densely hairy leaves and its relatively showy, yellowish flowers. Native in SE. Europe and Asia Minor.

A. speciosa Buser
Occasionally seen in gardens. Like *A. mollis* in flower colour and size, but with much more deeply-lobed leaves and rather ascending hairs on the stem and petiole. Native of the Caucasus.

A. tytthantha Juz.
Native of Crimea, but widely distributed in European Botanic Gardens, apparently as a 'weed'. Naturalized in S. Scotland, probably via the Royal Botanic Garden, Edinburgh. It can be distinguished by the following combination of characters: small flowers, leaves hairy on both sides and some downwardly-directed hairs on the stems or petioles.

S. M. WALTERS

CULTIVATED BRITISH PLANTS IN A NATURALISTIC GARDEN SETTING

Rock Garden Dept., Royal Horticultural Society's Garden, Wisley

In view of the fact that the part of the garden where, in general, the highest proportion of plants of British origin or parentage is grown is the water garden, and the boggy area around it, combined with the knowledge that water, and especially moving water, can be a major attraction in exhibits and shows, it was decided to design the exhibit around a small pool and waterfall.

In a corner site in the R.H.S. New Hall a pool of approximately 12 ft. by 18 ft. was constructed with a sheet of black butyl rubber over a wooden framework—allowing for a depth of only $4\frac{1}{2}$ in. of water. A carefully prefabricated water-fall was placed in one corner of the pool, with a small 'Otter' pump beneath it to re-circulate the water and provide ripples, splashes, and a trickling sound all day long.

The structure of the waterfall and the shape of the pool had to be hidden and made 'natural' by appropriate planting, and many aquatic plants were grown or fixed in containers shallow enough to be covered by the $4\frac{1}{2}$ in. of water. Thus the outline of the pool was 'rounded out' and the plants really looked as though they were growing in it. They did grow, too, and by the time the exhibit was dismantled on the afternoon of Wednesday, 6 September, it looked even more natural.

A background of Scots Pines and other trees and shrubs was arranged, with birches and willows marking the passage of the stream behind the fall.

A groundwork of heathers in flower and here and there a scattering of pine needles, with even a few fungi and tufts of moss, clothed the floor. Here, too, some line-drawings of early spring flowers showed what had come and gone.

Some cut herbaceous flowers staged in hidden vases, (*Eupatorium, Achillea, Ranunculus, Polygonum,* etc.) together with various grasses helped to merge the background with the Irises, Sagittarias, *Myosotis,* etc. overlapping the water's edge. Among these and out into the water ran *Menyanthes, Nymphoides* and the well established aliens, *Calla* and *Mimulus*, with clumps of *Acorus* (Sweet Flag). In the open water itself, beside the stepping-stones, were *Stratiotes* and *Hydrocharis* with a scattering of the inevitable species of *Lemna* and some *Azolla*. In the foreground, turf made an informal outline, and a 'frame for the picture'.

Between turf and pool, small outcrops of rock made homes for groups of mountain plants, seaside plants etc., and some wood-landers merged back into the trees.

A group of peat blocks made a site for the plants at home in acid soil. Bog plants were placed near the pool, with the smaller ones in the foreground. Together with *Sphagnum* moss these included *Anagallis tenella*, the leaves of *Viola palustris*, and turf from our alpine meadow full of that surprisingly extensive colonist, *Wahlenbergia hederacea*. Close by was *Gentiana pneumonanthe*.

In another corner were examples of fruits, vegetables, herbs and border plants of British descent.

The planting made a fascinating exercise in ecology. A very large number of species was used, with their variegated and other cultivars, but the surprising thing was that so very many more were assembled and *not* used, either because they did not quite fit in or because they had withered in the heat and drought.

The Committee and Council of the Royal Horticultural Society were kind enough to state that the exhibit was of Gold Medal standard.

W. K. ASLET

MENTHA

Living plants of a number of Mints were exhibited to illustrate and supplement the paper presented to the Conference. The plants represented in the exhibit were:

Mentha pulegium, M. suaveolens—wild type, *M. suaveolens*—variegated cultivar, *M. longifolia* subsp. *longifolia, M. spicata*—glabrous form, *M. spicata*—hairy form, *M. spicata*—rugose form, *M.* × *villosa*—glabrous form, *M.* × *villosa* nm. *alopecuroides, M. aquatica, M.* × *piperita, M.* × *piperita* (synthesised by author), *M.* × *piperita* nm. *citrata, M. arvensis, M.* × *gentilis, M.* × *gentilis* nm. *variegata, M.* × *smithiana*.

R. M. HARLEY

SNOWDROP PHOTOGRAPHS

This exhibit was staged to show the subtle differences which enable one to decide that a certain Snowdrop is sufficiently distinct to warrant a separate name, differences which can, in part, be illustrated by means of black and white photographs; in addition it had the aim of amplifying the lecture and presenting examples of the Snowdrops which had been mentioned, together with a few others. Diagnostic features included the leaves, the shapes and the green markings (if any) on both the inner and outer segments of the flower, together with the general appearance. If mild weather has prematurely pushed open the flowers, the foliage will often be undeveloped and not show the requisite features. On wet days there is usually insufficient light by which to photograph the plants, in addition, under these circumstances, the outer segments of the flower will cover the inner—there is a minimum temperature at which the flowers will open and this is not exceeded in wet winter weather; photographs taken then are useless. There is a short period before the flowers go over when they will stay open regardless of the temperature, but by then they are often very ragged or damaged. I find I must lie on the ground to obtain the right view! Finally, if the Snowdrop is being cultivated, it is not usually 'typical' until the third year after planting.

Many rude things have been written about labels being moved and the plant, therefore, bearing the wrong name. Sorting out black and white negatives or prints requires a reliable and competent recording system; I have photographs of at least one hundred and fifty distinct (to me) Snowdrops. One bulb which had never flowered for its two earlier owners, did this year; it had been cultivated and kept labelled for about fifty years before it flowered. The above illustrates some of the fun of producing such an exhibition.

Examples were given of 'wild' Snowdrops, both single and double *Galanthus nivalis* in Norfolk, Berkshire, Wiltshire and Cambridgeshire, showing natural habitat and a close-up of the flowers. Varieties of *G. nivalis* found in Great Britain and selected for cultivation were also shown. Naturalized clones of *G. elwesii* and *G.* 'Straffan' were illustrated. Natural wild hybrids of *G. elwesii* × *nivalis* rounded off the exhibit.

RICHARD NUTT

THE DISTRIBUTION OF MISTLETOE ON SELECTED HOSTS

Eight maps were exhibited showing the distribution of records of Mistletoe on the most frequent host plants. Five of these are reproduced as illustrations to the paper on p.139.

F. H. PERRING

BULBFIELD WEEDS IN THE ISLES OF SCILLY

A number of photographs of weeds in the Isles of Scilly bulbfields were exhibited.

The purpose of the exhibit was to draw attention to the unique assemblage of weed species which occur in the Isles of Scilly bulbfields. While the bulbs are dying down, the fields are left uncultivated, resulting in a prolific growth of weeds by the middle of May. This suits a number of thermophilic species which complete their life cycles early in the year. A number of very rare British native plants reach their greatest abundance in the Isles of Scilly, including *Briza minor, Polycarpon tetraphyllum, P. diphyllum, Fumaria occidentalis, Ornithopus pinnatus* and the hepatic *Sphaerocarpos texanus.* To these are added an increasing number of established aliens, including *Aira caryophyllea* subsp. *multiculmis, Ranunculus muricatus, R. marginatus, Gladiolus byzantinus, Allium roseum, Oxalis pes-caprae, O. articulata* and *Crassula decumbens.*

Of these weed species, only a few are really troublesome. Both the named *Oxalis* species are now very serious pests, reproducing rapidly from bulbils to choke all other vegetation. Herbicides only serve to stimulate bulbil production. *Gladiolus byzantinus,* an abundant relic of former cultivation, is also regarded with disfavour; the corms act as reservoirs of disease, which can then spread to more susceptible modern horticultural varieties.

Following the accepted continental system of naming plant communities, the bulbfield weeds fall somewhere between the weed association *Spergulo-Chrysanthemetum segeti* (Br. Bl. & De L.) Tx. 1937 of Britain and Western Europe, and the association *Fumarietum bastardii* Br. Bl. 1952 of W. Britain and Ireland. On sporadically cultivated coastal sands, the communities are referable to the alliance, *Thero-Airion,* which typically contains natural communities of open, sandy areas. However, the abundance of introduced, thermophilic species makes these Scilly bulbfield communities virtually unique.

Conservation of weed communities is something which is rarely suggested, although it is thought normal to conserve so-called 'natural' vegetation. On the mainland it is now difficult to find intact weed communities which could be conserved in any case, but the Isles of Scilly bulbfields are very different. Here we have an intact weed assemblage which, *Oxalis* spp. apart, is not of great economic importance. Early summer tourists certainly find the sheets of *Chrysanthemum segetum* and other species aesthetically pleasing. As already stated, the fields contain a number of nationally rare and very rare species, and the communities are virtually unique. Surely these are powerful arguments in favour of their conservation? Obviously conservation of weed communities poses a number of practical problems, but they are being solved in the Netherlands, and could be so, here.

A. J. SILVERSIDE
(University of Durham)

TEESDALE GENE BANK

Department of Botany, University of Durham.

The exhibition displayed, by means of panels of script and seven colour photographs, the *raison d'etre,* management and recording of the Teesdale Gene Bank established at Durham during 1970.

The twenty single-species populations being conserved were transplanted from the Cow Green area in Upper Teesdale shortly before the rising waters of the new reservoir would have caused their loss. They were planted in Teesdale soil, peat and sugar limestone in tubs made up from concrete sewage pipes and established at the Gene Bank in the University's new Botanic Garden. As such, they are able to provide both comparative data on growth and performance with the remaining populations in Teesdale and with a similar collection at a parallel gene bank at Manchester, and they also provide a stock pile of material for future research, thus eliminating the necessity for further reducing the remaining populations in Teesdale.

The populations are watered automatically, and are weeded regularly to attempt to reduce competition. The recording consists of collecting data on such aspects as numbers of shoots, percentage flowering and numbers of flowers and fruits per inflorescence. Now, during their second year, most of the species are thriving and are considerably larger than their counterparts on the Fell. These differences are of a phenotypic nature and the possibility of genetic change is minimised by collecting all the seed produced.

In addition three different turf communities were also transplanted and plants that were surplus to the gene bank requirement were planted along the banks of a 'Teesdale-simulated' stream into which Teesdale algae have been introduced.

The Gene Bank, together with the stream, is thus able to serve conservation, research, educational and amenity functions.

J. H. Gaman

Cultivars of British Native Trees and Shrubs

Messrs Hillier and Sons of Winchester

It is commonly assumed by the layman that the gardens of England are full of plants of foreign origin. So deeply is this idea embedded that he has been known to show surprise, even disbelief when faced with evidence to the contrary.

Nowhere is this misconception more apparent than in the field of trees and shrubs, where the average visitor to the garden, park or arboretum rarely finds his way beyond the flamboyant but ubiquitous Japanese Cherry. True, our native woody flora is modest compared with most temperate countries and pales to insignificance against the wealth of North America and Eastern Asia. It does, however, boast individual species which can at least hold their own with foreign competitors, trees such as the magnificent Beech (*Fagus sylvatica*), the billowy White Willow (*Salix alba*) and the stately English Elm *(Ulmus procera)*, now alas beleaguered. Our native shrubs too can muster several contenders for garden honours in the shape of the "double season" Guelder Rose (*Viburnum opulus*) the beloved May (*Crataegus* spp.) and the chrome-plated Gorse (*Ulex* spp.).

The English Holly (*Ilex aquifolium*) stands supreme in a genus of some four hundred species, as the most familiar and beloved of Christmas decorations.

However, it is not so much the species of native trees and shrubs which have established themselves in cultivation but the multitudinous forms and cultivars which they have collectively produced.

Ever since man began cultivating plants for ornamental purposes his eye has noted the odd and the unusual. This mania (as it became) for collecting and disseminating the curious and amusing deviations from the normal reached its peak during the 19th century, when the cultivars of Holly, Box and Yew alone numbered in the hundreds and became the backbone of Victorian gardening.

The early part of the present century witnessed a breaching of the doors to China and from this vast botanical treasure-chest flowed riches which assailed and eventually swamped the gardens of the west.

Soon our humble native species and their cortège of the curious, were neglected and forgotten as bright-berried *Berberis*, *Cotoneaster* and *Viburnum* burned their way into the English autumn, giant Magnolias threw pink chalices into the air, and a host of fiery Maples with alabaster-striped stems thronged our arboretums.

Gradually the incredible diversity of this latest influx was absorbed into an already polymorphic garden flora. Meanwhile the plantsman's insatiable appetite continued to expand, despite the advent of the small garden. Unusual yet useful forms of British native trees and shrubs were in demand again.

To this constant and continuing demand has now been added the needs of the New Town planner, the recreationalist, and the motorway builder, to whom a plant's utility value is often more important than its ornamental virtue.

Our native trees and shrubs, together with their cultivars, are horticulturally of the utmost importance and are used in most forms of gardening and landscaping. There are signs that this tendency is increasing, helped no doubt by a new general awareness of our native environment and its conservation.

The following is a list of British native trees and shrubs exhibited during the Conference. The exhibit was awarded the Lindley Silver-Gilt Medal by the Royal Horticultural Society.

Acer campestre 'Schwerinii'		*Buxus sempervirens* 'Argentea'	
Alnus glutinosa 'Aurea'		,,	,,	'Aurea Pendula'
,,	,,	'Imperialis'	,,	,,	'Aureovariegata'
,,	,,	'Incisa'	,,	,,	'Elegantissima'
,,	,,	'Laciniata'	,,	,,	'Handsworthensis'
,,	,,	'Pyramidalis'	,,	,,	'Hardwickensis'
Arbutus unedo 'Integerrima'		,,	,,	'Latifolia'
,,	,,	'Quercifolia'	,,	,,	'Latifolia Bullata'
Betula pendula 'Fastigiata'			
,,	,,	'Purpurea'	,,	,,	'Latifolia
Betula pubescens 'Crenata Nana'			Macrophylla'

Buxus sempervirens 'Latifolia
          Maculata'
,,   ,, 'Longifolia'
,,   ,, 'Marginata'
,,   ,, 'Myosotifolia'
,,   ,, 'Myrtifolia'
,,   ,, 'Pendula'
,,   ,, 'Prostrata'
,,   ,, 'Rosmarinifolia'
,,   ,, 'Rotundifolia'
,,   ,, 'Suffruticosa'
Carpinus betulus 'Aspleniifolia'
,,   ,, 'Fastigiata'
,,   ,, 'Pendula'
,,   ,, 'Variegata'
Corylus avellana 'Aurea'
,,   ,, 'Contorta'
,,   ,, 'Heterophylla'
Crataegus monogyna 'Pteridifolia'
,,   ,, 'Tortuosa'
Crataegus laevigata 'Gireoudii'
Euonymus europaeus 'Atropurpureus'
,,   ,, 'Aucubifolius'
Fagus sylvatica 'Albovariegata'
,,   ,, 'Cochleata'
,,   ,, 'Cockleshell'
,,   ,, 'Cristata'
,,   ,, 'Dawyk'
,,   ,, 'Grandidentata'
,,   ,, 'Heterophylla'
,,   ,, 'Latifolia'
,,   ,, 'Pendula'
,,   ,, 'Prince Georges of
          Crete'
,,   ,, 'Purpurea'
,,   ,, 'Purpurea Pendula'
,,   ,, 'Riversii'
,,   ,, 'Rohanii'
,,   ,, 'Roseomarginata'
,,   ,, 'Rotundifolia'
,,   ,, 'Zlatia'
Fraxinus excelsior 'Crispa'
,,   ,, 'Erosa'
,,   ,, 'Jaspidea'
,,   ,, 'Pendula'
Ilex aquifolium 'Angustifolia'
,,   ,, 'Argenteomarginata'
,,   ,, 'Argenteomarginata
          Pendula'
,,   ,, 'Argentea Mediopicta'
,,   ,, 'Aureomarginata'
,,   ,, 'Crassifolia'
,,   ,, 'Crispa'
,,   ,, 'Crispa Aureopicta'
,,   ,, 'Donningtonensis'
,,   ,, 'Elegantissima'
,,   ,, 'Ferox'
,,   ,, 'Ferox Argentea'
,,   ,, 'Ferox Aurea'
,,   ,, 'Flavescens'
,,   ,, 'Foxii'

Ilex aquifolium 'Golden Milkboy'
,,   ,, 'Grandis'
,,   ,, 'Handsworthensis'
,,   ,, 'Handsworth New
          Silver'
,,   ,, 'Hastata'
,,   ,, 'Heterophylla'
,,   ,, 'Heterophylla
          Aureomarginata'
,,   ,, 'Laurifolia'
,,   ,, 'Laurifolia Aurea'
,,   ,, 'Madame Briot'
,,   ,, 'Monstrosa'
,,   ,, 'Muricata'
,,   ,, 'Myrtifolia'
,,   ,, 'Myrtifolia Aurea'
,,   ,, 'Ovata'
,,   ,, 'Ovata Aurea'
,,   ,, 'Pendula'
,,   ,, 'Recurva'
,,   ,, 'Scotica'
,,   ,, 'Silver Queen'
,,   ,, 'Watererana'
Populus alba 'Richardii'
Populus tremula 'Pendula'
Prunus avium 'Decumana'
Prunus spinosa 'Purpurea'
Quercus petraea 'Laciniata'
,,   ,, 'Mespilifolia'
,,   ,, 'Purpurea'
Quercus robur 'Atropurpurea'
,,   ,, 'Concordia'
,,   ,, 'Cristata'
,,   ,, 'Cucculata'
,,   ,, 'Fastigiata'
,,   ,, 'Pendula'
,,   ,, 'Variegata'
Quercus × rosacea 'Pectinata'
Ribes alpinum 'Aureum'
,,   ,, 'Nanum'
Salix alba 'Chermesina'
,,   ,, 'Vitellina'
,,   ,, 'Sericea'
Salix cinerea 'Tricolor'
Sambucus niger 'Albovariegata'
,,   ,, 'Aurea'
,,   ,, 'Heterophylla'
,,   ,, 'Laciniata'
,,   ,, 'Pulverulenta'
,,   ,, 'Purpurea'
,,   ,, 'Pyramidalis'
Sorbus aria 'Chrysophylla'
,,   ,, 'Cyclophylla'
,,   ,, 'Decaisneana'
,,   ,, 'Lutescens'
,,   ,, 'Pendula'
,,   ,, 'Quercoides'
Sorbus aucuparia 'Aspleniifolia'
,,   ,, 'Beissneri'
,,   ,, 'Dirkenii'

Taxus baccata 'Adpressa'	*Taxus baccata* 'Variegata'
„ „ 'Adpressa Variegata'	„ „ 'Washingtonia'
„ „ 'Amersfoort'	*Tilia platyphyllos* 'Aurea'
„ „ 'Aurea'	„ „ 'Laciniata'
„ „ 'Cheshuntensis'	„ „ 'Rubra'
„ „ 'Dovastoniana'	*Ulmus glabra* 'Camperdownii'
„ „ 'Dovastonii Aurea'	„ „ 'Exoniensis'
„ „ 'Elegantissima'	„ „ 'Lutescens'
„ „ 'Ericoides'	„ „ 'Pendula'
„ „ 'Fastigiata'	*Ulmus procera* 'Argenteovariegata'
„ „ 'Fastigiata	„ „ 'Louis van Houtte'
Aureomarginata'	„ „ 'Silvery Gem'
„ „ 'Glauca'	*Viburnum opulus* 'Aureum'
„ „ Repandens'	„ „ 'Compactum'
„ „ 'Repens Aurea'	„ „ 'Notcutt's Variety'
„ „ 'Semperaurea'	„ „ 'Xanthocarpum'
„ „ 'Standishii'	

ROY LANCASTER

PAINTINGS OF BRITISH PLANTS IN CULTIVATION

Paintings of a number of British plants (native and alien), made for pleasure, mainly during the years 1964 to 1972, were exhibited. Although they were all painted from cultivation in my garden the plants or seed came from many sources. The plants illustrated in the exhibit were:

Acanthus spinosus
Acer platanoides
Ajuga genevensis
Alchemilla confusa
Amelanchier confusa
Anagallis arvensis var. caerulea
Anemone blanda
Aponogeton distachyos
Argemone mexicana
Asperula arvensis
Blechnum spicant
Carex atrata
Chenopodium polyspermum
Cucubalus baccifer
Datura stramonium
Datura tatula
Drosera rotundifolia
Festuca vivipara
Geranium phaeum
Geranium pratense
Geum rivale
Helleborus foetidus
Hieracium brunneocroceum
Homogyne alpina
Hydrocharis morsus-ranae

Hypericum calycinum
Iris foetidissima
Iris pseudacorus
Lagurus ovatus
Lamium album
Lilium pyrenaicum
Ludwigia palustris
Narcissus obvallaris
Nicandra physalodes
Nymphoides peltata
Oenothera stricta
Ornithogalum umbellatum
Oxalis corniculata var.
    atropurpurea
Oxalis megalorrhiza
Oxytropis halleri
Pinguicula grandiflora
Poa bulbosa var. vivipara
Polypodium vulgare
Potamogeton acutifolius
Ranunculus circinatus
Sagittaria rigida
Sarracenia purpurea
Silene maritima

BARBARA EVERARD

SPECIMENS OF PLANTS RELEVANT TO THE CONFERENCE

A small exhibit of herbarium specimens was displayed illustrating points relevant to the Conference.

D. MCCLINTOCK

APPENDIX II

THE EXCURSION TO BOX HILL AND THE ROYAL HORTICULTURAL SOCIETY'S GARDEN, WISLEY

As part of the Conference a day excursion was arranged for a visit to Wisley Garden and to the chalk flora of Box Hill, both in Surrey. The pleasant weather helped to make the excursion a success and it was so well attended, with about 120 participants, that the group was divided into two parties, one half going to Box Hill in the morning and Wisley in the afternoon, while the other half reversed the order. However, the two parties followed the same routes, and the notes which follow are equally applicable to them both.

Box Hill

The excursion to Box Hill began from the Juniper Hall Field Centre by kind permission of the Warden, Mr John Sankey. From the Centre the parties made their way up Headley Lane and then to the foot of the Dogwood slope to Juniper Top. Members were surprised by the small size of the Clustered Bellflower, *Campanula glomerata,* growing on this shallow soil. There were not many flowering plants still in good flowering condition but it was the ideal time of year to see the Autumn Lady's Tresses, *Spiranthes spiralis,* which was producing numerous scattered spikes amongst the short turf between the main path and the Yew wood. Search was made in vain for the Ground-pine, *Ajuga chamaepitys,* on the bare patches of chalky soil just outside this wood on the west slope of the ridge. It was still possible to demonstrate the fallen skeletons of the Junipers, *Juniperus communis,* under the spreading branches of the Yew, *Taxus baccata.* These Junipers had given the ridge and valley their name before being shaded out, but it was very difficult to find the two or three Junipers, now in very poor condition, remaining in the open *Brachypodium*-grassland between the blocks of Yew wood on this slope.

Approaching the top of the ridge the visitors were impressed by a change in the vegetation, from the chalk grassland plants, e.g. Upright Brome (*Zerna erecta*), Viper's Bugloss (*Echium vulgare*), Yellow-wort (*Blackstonia perfoliata*) and Wild Basil (*Clinopodium vulgare*), to those more reminiscent of acid heathland–Wood-sage (*Teucrium scorodonia*), Bracken (*Pteridium aquilinum*), Heather (*Calluna vulgaris*) and Bell Heather (*Erica cinerea*), amongst Bent-grasses (*Agrostis* spp.), where the Tertiary sands and gravels form a cap on the plateau top and get washed down over the underlying chalk and clay with flints.

The parties walked through the Oak and Birch wood on the plateau and down through the fine Beech wood on the thin chalk soil below. Beside the path under the Beech trees there were several fine plants of the Broad Helleborine, *Epipactis helleborine*. Continuing along this path to the top of a grass slope they were able to see several plants of the Mountain St. John's Wort, *Hypericum montanum*, most were in fruit but those in the shade of Hawthorn had some flowers. Lower down the slope to Juniper Bottom, the Hairy St. John's Wort, *Hypericum hirsutum*, was passed and also Ploughman's Spikenard, *Inula conyza*.

The return route along Juniper Bottom took the parties past Buckthorn, *Rhamnus catharticus*, and Box, *Buxus sempervirens*, as components of the chalk scrub, and Vervain, *Verbena officinalis*, which has been growing in the same sites amongst grasses beside the path for at least the last twenty years.

WISLEY GARDEN

Although time was restricted the parties were shown around the Garden and its various collections by members of the staff. Amongst the many plants which attracted attention were the semi-naturalized clumps of Willow Gentians (*Gentiana asclepiadea*) in the Wild Garden, a sward of the Ivy Campanula (*Wahlenbergia hederacea*) on the Alpine Meadow and flourishing colonies of the peloric-flowered form of the Common Toadflax (*Linaria vulgaris*) in the Alpine Department. Much interest was also shown in the trial of Perennial Ornamental Grasses, which contained a number of desirable forms of several British species. Lunch, was eaten on the restaurant terrace overlooking the attractively landscaped area of the garden known as Seven Acres.

A small exhibit of photographs and literature illustrating specific aspects of British botany was staged in the laboratory.

P. J. WANSTALL & J. E. ELSLEY

APPENDIX III

THE SPECIES OF ACAENA WITH SPHERICAL HEADS CULTIVATED AND NATURALIZED IN THE BRITISH ISLES

P. F. YEO

University Botanic Garden, Cambridge

This review concerns the species of *Acaena*, naturalized or cultivated in Britain, with the flowers gathered into capitula which are spherical or nearly so and which have 0–4 spines at the top of the receptacle and either one achene (Sect. *Ancistrum* (J. R. & G. Forst.) DC. and Sect. *Anoplocephala* Citerne) or two achenes (Sect. *Microphyllae* Bitter). (The sectional names of the latest monograph (Bitter, 1910–1911) cannot be adopted in full because they are in conflict with the International Code of Botanical Nomenclature).

Bitter's monograph (1910–1911) covers the variation displayed by the genus *Acaena* in great detail, but by its very elaborateness it is rendered difficult to use. There seems to be a marked inequality in the variation allowed within species, some of them being very wide and embracing numerous subspecies, others being narrow and having the appearance of only minor variants.

Recently, the 'Flora of New Zealand' (Allan, 1961) has furnished an excellent revision for that country, recognizing 15 species, and a revision by Grondona (1964) covers part of South America, but unfortunately does not attempt to represent the variation of the *A. magellanica* group.

The key which I present below is based entirely on vegetative characters; any specimen to which this key is applied should be checked with the descriptions, and if there are substantial discrepancies it will have to be assumed that the specimen does not represent one of the species here dealt with. There is striking diversity in the colour and texture of the leaflets in *Acaena* but in order to make the key easy to use with dried material I have employed these variations as little as possible. When collecting specimens to be dried it is very helpful to *remove carefully a few leaves, complete with stipules, and press them separately, to note the leaf-colour, and to break up some capitula.*

The descriptions are based chiefly on British and cultivated material, and are accompanied by a selected synonymy and observations on the history and occurrence of the species. My records of British naturalized occurrences are certainly incomplete, and a full compilation is desirable.

Acaena L., Mantissa Plantarum 2: 200 (1771).

Undershrubs with creeping and rooting stems or occasionally rhizomes, these stems producing axillary shoots which are usually capable of flowering in their second year. Leaves pinnate, stipulate; leaflets usually toothed, asymmetric at the base. Inflorescence (in our species) terminal on main and lateral shoots. Flowers hermaphrodite or gynodioecious. Receptacle forming a cupule, usually armed with barbed spines at the fruiting stage. Sepals usually 4; petals none; stamens 1–7 (2–4 in our species); carpels 1 or 2 (in our species).

A genus of wide distribution in the Southern Hemisphere, extending to California and the Hawaiian Islands. In Britain sometimes naturalized, probably occasionally escaping from gardens, but more usually accidentally introduced, especially with wool shoddy.

The apparent extreme inequality of Bitter's specific concept has been mentioned on p. 193, and it is when dealing with the South American species that he adopts a narrow species concept; his Monograph (Bitter, 1910–1911) added considerably to the number of species in what I call the *magellanica* group. Grondona (1964), on the other hand, went to the opposite extreme, and his *A. magellanica* includes over 30 binomials as synonyms. While the taxonomic problem is undoubtedly difficult, the morphological range which these binomials represent is so enormous that one feels it ought to be possible to do something better than just lump the lot, and the way in which this might be done is suggested on p.198. The group is represented in this paper by the first four taxa, and fortunately these are quite distinct from one another in their naturalized or horticultural populations, and can all be identified with particular described taxa. Of these four, *A. macrostemon* has no nomenclatural problems. For the others, which I call *A. magellanica* subsp. *magellanica* (1), *A. magellanica* subsp. *laevigata* (2) and *A. affinis* (3), the essential citations are set out and discussed below.

KEY TO SELECTED SPECIES OF ACAENA

In this key the phrase 'distal leaflets' applies to the *two* distal pairs of lateral leaflets. The 'adnate portion of the stipule' is that which is adnate to the petiole.

1. Distal leaflets at least 1⅓ times as long as broad and some of them normally 1½–2½ times as long as broad:

   2. Adnate portion of stipule 15–18mm long; leaflets glaucous above .. .. 4. **macrostemon**

   2. Adnate portion of stipule 2–5mm long; leaflets not glaucous above:

   3. Distal leaflets with (12–)17–23 teeth   8. **ovalifolia**

   3. Distal leaflets with 8–12(–13) teeth:

      4. Leaflets bright glossy green above, other vegetative parts sometimes red-flushed; distal leaflets just less than twice, to more than $2\frac{1}{2}$ times, as long as broad .. .. 9. **novae-zelandiae**

      4. Distal leaflets light matt green above, edged and veined with brown; proximal leaflets above, rachises, petioles and stems, brown; distal leaflets not more than twice as long as broad .. 10. **anserinifolia**

1. Distal leaflets at most $1\frac{1}{2}$ times as long as broad:

   5. Leaves not glaucous above (use lens for dried material):

      6. Stipule-lobes 1 or 2:

         7. Leaflets flushed brown above, the distal not more than 4·5mm long, with 3–7 teeth .. 13. **microphylla**

         7. Leaflets deep green above, the distal 5·5–16mm long, with 7–9(–10) teeth 2. **magellanica** subsp. **laevigata**

      6. Stipule-lobes 3 or more in at least some of the leaves:

         8. Plant quite glabrous (except for bracts on scape); adnate portion of stipule 4–6mm long 6. **glabra**

         8. Plant not glabrous; adnate portion of stipule not more than 2mm long .. .. .. 11. **pusilla**

   5. Leaves glaucous above (use lens for dried material):

      9. Leaflets 7–9; plant densely hairy 12. **caesiiglauca**

      9. Leaflets 11 or more in the majority of leaves:

         10. Distal leaflets of most leaves more than 10mm long .. .. 3. **affinis**

         10. Distal leaflets of most leaves not more than 10mm long:

            11. Stipule-lobe 1 or 0:

               12. Adnate portion of stipule 1·5–2·5mm long; distal leaflets as wide as long 14. **inermis**

               12. Adnate portion of stipule 3mm long or more; at least some of the distal leaflets slightly longer than wide:

                  13. Distal leaflets with 3–7(–8) teeth; teeth not penicillate; woody stems stout, 2–7mm thick

                        1. **magellanica** subsp. **magellanica**

                  13. Distal leaflets with (5–)6–12 teeth; teeth penicillate (i.e. with an apical tuft of hairs); woody stems usually not more than 2mm thick

                        15. **buchananii**

11. Stipule-lobes 2–6 in at least some of the leaves:
    14. Leaflet-teeth penicillate (i.e. with an apical tuft
        of hairs)–sometimes weakly so; sinuses between
        teeth of distal leaflets extending $\frac{1}{4}$–$\frac{1}{3}$ of the way
        towards the midrib .. .. .. 7. fissistipula
    14. Leaflet-teeth not penicillate; sinuses between
        teeth of distal leaflets extending $\frac{2}{5}$–$\frac{1}{2}$($-\frac{3}{5}$) of the
        way towards the midrib .. .. 5. 'BLUE HAZE'

DESCRIPTIONS OF THE SPECIES

1. **Acaena magellanica** (Lam.) Vahl, Enum. Pl. 1: 297 (1804) subsp. **magellanica.**

Ancistrum magellanicum Lam., Tabl. Encycl. Méth. Bot. 1: 76 (1791), excl. 'β'. LECTOTYPUS: Ancistrum L. no. 15, Suppt. Du Détroit de Magellan. C[ommers]on. Lui. no. 272 (P).

Acaena venulosa Griseb. in Abh. Kön. Wiss. Gött. 6: 118 (1854) (System. Bemerk.). TYPUS: *W. Lechler,* pl. Magell. ed. R. T. Hohen., 978c, in collibus prope Sandy Point, Octobr. mens. (UPS, syntypus!).

Acaena glaucophylla Bitter in Biblioth. Bot. 17 (74): 155 & t. 16a (1910).

Acaena magellanica (Lam.) Vahl subsp. *venulosa* (Griseb.) Bitter, *op. cit.*: 168 (1910).

Acaena laevigata auctt. Brit.

Acaena adscendens auctt. Brit.

Woody stems up to 7mm thick; *herbaceous stems* pale green, flushed deep pink, glabrous; sterile axillary shoots rosetted. *Leaves* 2–4·5(–8·5) cm long; *stipule* with adnate portion 5–10mm long, foliaceous portion much shorter, entire or absent; rachis without subsidiary leaflets; *leaflets* 11–13(–17), upper surface light glaucous grey-green with impressed veins, glabrous, lower surface paler with prominent veins, glabrous; the two *distal pairs of leaflets* more or less flabelliform, 1–1$\frac{1}{4}$(–1$\frac{1}{3}$) times as long as broad (or longer when very luxuriant), 3·5–8(–12)mm long, 3·5–7(–10)mm wide, proximal margin sometimes slightly decurrent, the *teeth* 3–7(–8), subacute, not penicillate, clefts extending $\frac{1}{3}$–$\frac{1}{2}$ of the way towards the midrib. *Scapes* ascending, pale green, flushed purplish red, (5–)7–12(–17)cm long in fruit, glabrous; *capitula* spherical or very slightly longer than broad, 7–11mm long in flower, 18–24mm in fruit (including spines); *stamens* 2–4, dark red; *stigma* one, dark red, about as long as broad; *ripe cupules* c. 3·5mm long, with 4 bladdery and tubercled keels decurrent from the spines, glabrous or with fine reflexed hairs distally; *spines* 2–6·5mm long, barbed; minute supplementary barbed spines sometimes present between the main ones; subsidiary capitula well developed. Fig. 1.

Native in Southern Patagonia (Chile, Argentina). Introduced. V.C. 30: Maulden, 1963 (cult. 1964), *J. E. Lousley* W. 2217 (BM, K). V.C. 79: Galafoot, (cult. 1912), *J. Fraser* (K); Galashiels, 1963 (cult. 1965), *M. McCallum Webster* 9068 (BM)—form with extra large, elongated leaflets (see remarks below), & 1963 (cult. 1965), 9069 (CGE). V.C. 83, Levenhall, 1914 (cult. 1923), *J. Cryer* (BM, CGE, K).

I have seen the microfiche of the types of *Ancistrum magellanicum* Lam. and *A. magellanicum* β Lam. in the Lamarck Herbarium (P). The first of these two bears the words cited above and in addition two more entries in different hands: (1) Poterium?/foliis incisis spica floriferior/Cylindricis; fructiferis subrotundis spinulosis; (2) ancistrum magellanicum. lam. Entry (1) appears to be in the same hand as the locality.

The protologue of *A. magellanicum,* excluding the var. β, reads as follows: "T. 22, f. 1. A. foliolis ovatis inciso-pinnatifidis, spica capitato globosa. E Magellania. Commers." The locality of the herbarium specimen, the collector and the "foliis incisis" of the label clearly connect it with this protologue and make it a virtually obligatory lectotype for *Ancistrum magellanicum.* The second element included in *A. magellanicum,* the var. β, was cited by various later authors and is considered under my taxon 2 (below).

FIGURE 1

A. magellanica subsp. *magellanica* (66–60 Utrecht), × ½. (The Cambridge Botanic Garden accession number is given for every sample in Figs. 1–13).

The drawing in plate 22 accompanying Lamarck's description is extremely crude but it was probably made from material of var. β. However, this could not conceivably justify the reversal of the typifications which I propose here. The type specimen is very clearly the same as this taxon 1; the largest leaves are 2·5–3·5cm long and 8–11mm wide, the leaflets are nearly as broad as long with the teeth large in relation to the size of the leaflet, and one distal leaflet can be seen to have 3 teeth on the distal side, suggesting 7 or 8 teeth in all, while the proximal leaflets are often only 3-toothed.

The new description which Vahl gave when transferring *Ancistrum magellanicum* to *Acaena* is sufficiently detailed to show that he saw material; in saying that it resembles *Acaena trifida*, that it has trifid intermediate leaflets, and that its leaflets are ¼ inch (7mm) long he is evidently excluding our taxon 2 (below). Bitter doubtless saw Vahl's material (as he did in the case of *A. adscendens*) and he identified *A. venulosa* with it, saying "*A. venulosa* Gris. proves to be identical with *A. magellanica* Vahl in the narrower sense". The type number of *A. venulosa* represents this taxon, 1. (Curiously, Bitter nowhere mentions *Ancistrum magellanicum* Lam., as far as I have been able to find out.) Under present rules the subspecific epithet *venulosa* used by Bitter must be replaced by *magellanica*.

This taxon corresponds to Bitter's *A. magellanica* subsp. *venulosa* but also includes his *A. glaucophylla* described from cultivated plants which he suspected had been introduced by Dusén from Patagonia or Tierra del Fuego. It seems likely that the *A. magellanica* group can be broken down into three main polymorphic species, *A. magellanica, A. affinis* and *A. macrostemon. A. magellanica* subsp. *magellanica* itself will probably have to be defined eventually as comprising not only *A. glaucophylla,* with the subsidiary spines on the cupule, but also hairy forms such as those described as species by Bitter (1910: 160) under the names *A. oligodonta* and *A. floribunda* and placed near his *A. glaucophylla.* I have assumed that the abnormalities in Miss McCallum Webster's cultivated specimen of the plant from Galashiels are due to unusual luxuriance; however, in their leaflet shape and in the possession of subsidiary capitula they resemble *A. acroglochin* Bitter(1910: 163), though this too may not be specifically distinct from *A. magellanica.* (Somewhat elongated leaflets are seen in the two shade-leaves at the left of Fig. 1).

2.   **Acaena magellanica** (Lam.) Vahl susp. **laevigata** (Aiton f.) Bitter in Biblioth. Bot. 17 (74): 170 (1910).

Ancistrum magellanicum β Lam., Tabl. Encycl. Méth. Bot. 1: 76 (1791).
Acaena adscendens Vahl, Enum. Pl. 1: 297 (1804). LECTOTYPUS: Poterium humile Jus. Poterium sanguisorba inodora. *Herb. Lamarck* (P).

Ancistrum magellanicum β *humile* Pers., Syn. Pl. 1: 30 (1805).
Acaena laevigata Aiton f., Hort. Kew. ed. 2, 1: 68 (1810).
LECTOTYPUS: Acaena laevigata, [cult.] Hort. Kew. 1792 (BM).
Acaena magellanica sensu Hook. f., Fl. Antarct. 2: 268 (1846).

Woody stems up to 10mm thick; *herbaceous stems* green, sometimes flushed pinkish red, glabrous; sterile axillary shoots rosetted. *Leaves* 3–9·5cm long, very sparingly glandular-hairy; adnate portion of *stipule* 4–15mm long, foliaceous portion much shorter, entire or 2-lobed; rachis without small subsidiary leaflets; *leaflets* 9–15, thickish, upper surface rather dark green, not obviously glaucous though with a slight removable bloom, glabrous, lower surface glaucous green with conspicuously reticulate veins, glabrous or with a few fine hairs; the two *distal pairs of leaflets* oblong or nearly square in small leaves, cuneate-obovate in larger leaves,$(1-)1\frac{1}{4}-$

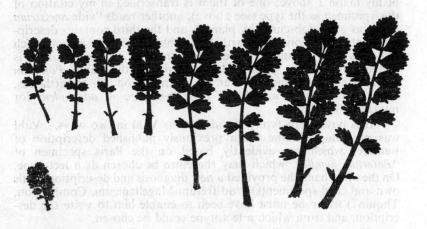

FIGURE 2
A. magellanica subsp. *laevigata* (46–69 Bonn), × ½.

$1\frac{1}{3}(-1\frac{1}{2})$ times as long as broad, 5·5–16mm long, 4–10·5mm wide, proximal margin rarely slightly decurrent, the *teeth* 7–9(–11), obtuse to subacute, not penicillate, clefts extending $\frac{1}{4}-\frac{1}{3}$ of the way towards the midrib. *Scapes* erect or ascending, green or flushed reddish, (6–)10–17cm long in fruit, pilose, especially above; *capitula* slightly elongated, 12–15mm long and 11–12mm wide in flower, 21–24mm long, and nearly spherical, in fruit (including spines); *stamens* 2, dark red; *stigma* one, dark red, about as long as broad; *ripe cupules* c. 3·5mm long, narrowed at base, bladdery above, ribbed, pilose distally; *spines* 2–4, c. 3mm long, red distally, barbed; subsidiary capitula well developed. Fig. 2.

Native in the Falkland Islands.

(Although I have not seen any specimens of this subspecies except from the Falkland Islands, it was, as I show below, in cultivation in Paris in 1790 or earlier. The record of botanical collecting in the Falklands (Moore, 1968) indicates that this material could hardly have been collected there, which suggests that the plant may occur also in the Magellan Region).

Acaena adscendens Vahl was published with a diagnosis, a description and the citation "Ancistrum magellanicum β foliolis latioribus serratis. Lamarck illustr. 1. p. 76. Habitat ad fretum Magellanicum. Commerson, Thouin". This is a quite definite citation of Lamarck's *Ancistrum magellanicum* β, which was published as follows: "β Idem, foliolis latioribus serratis. Poterium humile. H. R."

The microfiche of the specimen of var. β in the Lamarck Herbarium (P) shows three labels in the same hand as 'entry no. (2)' of my taxon 1 above; one of them is transcribed in my citation of this specimen as the type (see above), another reads "vide *ancistrum* an hujus g. . . . . [obscured by plant]", and the third contains descriptive matter not connected with the published diagnosis, but which indicates that the stamens are two in number. This specimen is connected with the published diagnosis by the name *Poterium humile,* and it is in fact an example of this taxon 2. Nevertheless, Hooker (*l.c.* above) and Bitter used the name *A. adscendens* for my taxon 3 (below).

It is possible to typify *A. adscendens* Vahl in two ways. Vahl was providing a name for a previously published description of Lamarck which was evidently based on the Paris specimen of "*Poterium humile*", which may therefore be chosen as a lectotype. On the other hand, he provided a new diagnosis and description of his own and cited specimen(s) ("ad fretum Magellanicum. Commerson, Thouin") which he must have seen to enable him to write the description, and from which a lectotype could be chosen.

The specimens of *Acaena adscendens* in Vahl's herbarium (C) were discussed by Bitter (1910: 176). There were two labelled *Acaena adscendens* in Vahl's hand, and according to Bitter one was a scanty specimen labelled "leg. Commerson a fretu magell. ded. Dr. Thouin", which was referable to *A. adscendens* sensu Hook. f., that is, taxon 3 (below), and the other was labelled "Schum." (and so presumably from the herbarium of, but not collected by, the Danish botanist H. C. F. Schumacher), and was identical with *A. laevigata* Aiton f., that is, this taxon 2. If the first of these specimens was correctly identified by Bitter and could be chosen as a lectotype it would be possible to continue using the name *A. adscendens* in its traditional sense. Bitter says that the leaflets of this specimen were pilose on the undersurface, and not merely on the midrib, so his determination can hardly be questioned. In that case, however, the labelling should perhaps be questioned, for Vahl

wrote "foliolis glabriusculis. . . . costa pilosiuscula".   Unfortunately,
the Vahl specimens were included in a parcel of 105 specimens sent
from Copenhagen in 1957 to San Isidro, Argentina, which never
reached its destination and must be presumed lost (information

FIGURE 3
A. affinis (354–61 Fish), ×1.

kindly supplied by Dr A. Skovsted). The Thus any doubts about the Vahl specimens are not likely to be cleared up, and the rejection of the Paris specimen as a lectotype could only be justified if Vahl's own description appeared to rule it out and to point to another species. The application of the name *A. adscendens* would then be determined by the description in the absence of a type.

In fact there are no points in Vahl's description of *A. adscendens* telling against its being this taxon 2, while this identification is supported by his describing the stems as decumbent, the plant as 7 inches (18cm) high, and the leaflets as numbering up to 15 and being obtusely serrate, veiny, and pilose on the midrib. On the other hand there is only one point in favour of identifying *A. adscendens* with taxon 3 (below), namely that the leaves are not described as clustered at the ends of the branches as in Vahl's description of *A. magellanica,* despite the fact that taxa 1 and 2 are similar in this respect. The explanation of this is doubtless that Vahl was dealing with somewhat lax cultivated material (the evidence for this statement is given below). Against *A. adscendens* being taxon 3 are the number of leaflets (though Bitter says this number *can* occur), the restriction of leaflet pilosity to the midribs, the obtuseness of the serrations, and the absence of mention of the blue colour or any glaucousness.

It would thus in my opinion be a departure from the advice of Recommendation 7b of the International Code of Botanical Nomenclature, 1972, to apply the name *A. adscendens* to taxon 3. The description points to this taxon 2, and the Lamarck specimen is thus a suitable lectotype. *Acaena adscendens* is, then, the correct name at specific rank for taxon 2, but the confusing situation which would result from this application of the name is avoided if the taxon is treated as a subspecies, in which case the nomenclature of Bitter's Monograph applies.

The evidence that the Paris specimen labelled "*Poterium humile*" was cultivated is as follows: (1) there is no stout woody stem, suggesting that the plant was still young and that it might have been difficult to collect woody material without destroying the plant, (2) the specimen is rather luxuriant and lax compared with normal wild material of this taxon, (3) it comprises flowering and fruiting heads and a separately mounted leaf, showing a degree of care in collection more likely to be achieved in a garden than on an expedition. That Lamarck was in fact citing a cultivated specimen in his published account of *Ancistrum magellanicum* β is shown by the letters "H. R." which he included, and which doubtless stand for "Hortus Regis." Not only did Vahl cite this name but so did Aiton. In addition, both authors felt they knew enough about the plant to treat it as a species separate from *Acaena magellanica,* despite Lamarck's scanty description, and both authors mention Thouin, Aiton saying it was he who introduced it into cultivation in Britain in 1790. André Thouin (1747–1824) was head gardener of the Jardin du Roi, Paris, and it must have been on his authority that Aiton and Vahl relied for the knowledge that the plant in cultivation which they received from him was var. β. I have examined Aiton's specimen cultivated at Kew in 1792 (BM) and it is clearly taxon 2. This is circumstantial evidence in support of my determination of the Paris specimen seen only in microfiche form. (I have thought it better to choose this as the lectotype of *A. laevigata*, rather than the Paris specimen). It is interesting to note that the name *A. adscendens* continued to be applied to this taxon in botanic gardens right up to the time of Bitter's Monograph (Bitter, 1910–1911: 171).

3. **Acaena affinis** Hook. f., Fl. Antarct. 2: 268 (1847). LECTOTYPUS: Christmas Habour, Kerguelen Island, on marshy low grounds like *Comarum palustre*, May 1840, Antarct. Exp. 1839–1843, *J. D. H.* 764 (K).

> *Ancistrum decumbens* Gaertner, Fruct. Sem. Pl. 1: 163 (1788) (teste D. W. H. Walton, 1971, vide infra).
> *Acaena adscendens* sensu Hook. f., *op. cit.* 267 & t. 96, (1847), sensu Vallentin & Cotton, Ill. Fl. Pl. Ferns Falkland Is., t. 17 (1921), & sensu Bitter in Biblioth. Bot. 17 (74): 175 (1910).
> *Acaena distichophylla* Bitter, in *op. cit.* 208, t. 20 & fig. 53 (1910).
> *Acaena decumbens* (Gaertner) D. W. H. Walton in Br. Antarct. Surv. Bull. 25: 30 (1971), non *A. decumbens* (L.f.) Druce (*Agrimonia decumbens* L.f.).

Woody stems up to 6mm thick; *herbaceous stems* sometimes creeping underground, otherwise bright pink, glabrous; sterile axillary shoots elongate, their leaves sometimes distichous.   *Leaves* (3–)6–12(–15)cm long (apart from greatly reduced ones at stem-bases); adnate portion of *stipule* (6–)9–13mm long, foliaceous portion much shorter, entire or 2–3-lobed; rachis sometimes with small subsidiary leaflets; *leaflets* 9–13, upper surface light bluish or greyish glaucous green, glabrous or pilose at the margins, lower surface grey-green, pilose, with the veins conspicuous; the two *distal pairs of leaflets* broadly obovate, oblong or suborbicular, 1–1½ times as long as broad, (6–)9–18mm long, (4·5–)7–16mm wide, proximal margin usually decurrent for up to 4mm, the *teeth* 9–14, acute or subacute, clefts extending ⅛–¼(–⅓) of the way towards the midrib.   *Scapes* ascending, red, 11–21cm long in fruit, appressed-pilose above; *capitula* spherical, 12–13mm diam. in flower, 20–25mm in fruit (including spines); *stamens* usually 4, dark red; *stigma* one, dark red, c. 6 times as long as broad; *ripe cupules* 4–5·5mm long, 4-ribbed or 2-ribbed and with 2 bladdery wings alternating with the ribs, pilose at apex; *spines* 4, up to 6 or 8mm long, red, barbed; minute supplementary spines sometimes present between the main ones. Fig. 3.

Native in the Magellan Region, Falkland Islands, South Georgia, Marion Island (Prince Edward Islands), Crozet Islands, Kerguelen Islands, Macquarie Island.

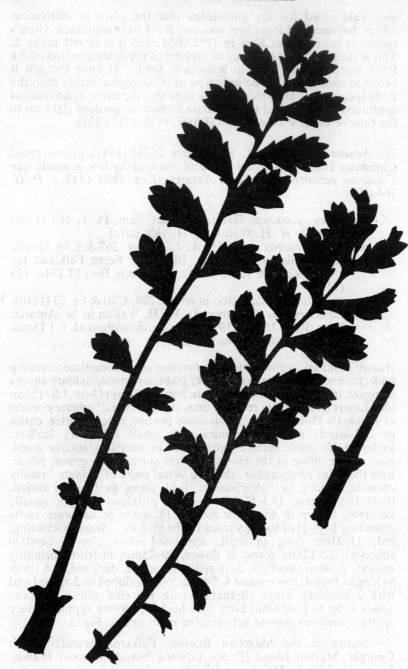

FIGURE 4
A. macrostemon (83–98 Ellacombe), ×1.

This taxon has traditionally been called *A. adscendens* but I have shown that this name belongs to taxon 2 (above). However, Bitter (1910: 179) considered that *A. affinis* was conspecific with *A. adscendens*, which he interpreted in Hooker's sense. Hooker distinguished *A. affinis* purely on account of the small recurved teeth on the cupule above and between the bases of the spines. Bitter found that these processes were not always recurved and in any case were not present in all Kerguelen material. He regarded them as vestiges of accessory spines which are still present and glochidiate in more primitive *Ancistra*, and considered that they did not entitle their possessors to specific separation. I have found the teeth very inconspicuously present in the cupules from one head on the lectotype sheet and more or less absent on another. In another Kerguelen specimen these teeth seemed to be quite absent (1874, *Moseley* (K)), while a specimen from Marion Island (*Bloemfontein University* 8562, coll. *Bakker* (K)) has a quite long slender tooth emerging between each pair of spines. I agree with Bitter's taxonomic conclusion here, and thus regard *A. affinis* as the earliest name available for this taxon. On the other hand it seems to me that some of Bitter's own new species assigned to his Subsections *Subantarcticae* and *Distichophyllae* are also synonyms of this taxon. In fact, the stock of this species received at the Cambridge Botanic Garden from Mrs Fish (see p. 51) appears to be identical with the plant illustrated by Bitter (1910: *l. c.* above) as *A. distichophylla*. A curious feature of the cupules of a plant raised in 1969 from seed collected in 1968 in King Edward Cove, South Georgia (see p. 51) is that the spines arise well below the apex. In South Georgia *A. affinis* hybridizes with *A. tenera* Alboff (Walton & Greene, 1971). It is not known to have become naturalized in Britain.

4.   **Acaena macrostemon** Hook. f., Fl. Antarct. 2: 269 (1847).

Woody stems up to 6mm thick, hardly persisting above ground; *herbaceous stems* sometimes creeping underground, otherwise pale green, flushed pink, sparsely and finely appressed-pilose; axillary shoots elongate. *Leaves* 6–12cm long (apart from greatly reduced ones at stem-bases); adnate portion of *stipule* 15–18mm long, foliaceous portion much shorter, entire; some of the rachises with small subsidiary leaflets; *leaflets* 11–19, more or less folded longitudinally, upper surface rather pale glaucous green, finely pilose, lower surface paler, rather thinly silky-hairy; the two *distal pairs of leaflets* obovate, oblong or lanceolate, c. $1\frac{1}{2}$ to nearly twice as long as broad (when flattened), 11–18(–22)mm long, 5·5–10(–12·5)mm wide (when flattened), proximal margin of one or both pairs usually decurrent for 1–6mm, the *teeth* 5–8, acute, not penicillate, clefts extending $\frac{1}{2}$-way or more towards the midrib. *Scapes* erect or ascending, pale green, villous, 18–30cm long in fruit; *capitula* spherical, 12–14mm diam. in flower, 23–28mm in fruit (including spines); *stamens* (2–)4; *stigma* one, deep pink, c. 4 times as long as broad; *ripe cupules* 4·5mm long, 2-winged, pilose distally; *spines* 4, 3–7mm long, straw-coloured, sometimes tinged pink distally, barbed. Fig. 4.

Native in Argentina and Chile.

The above description is based on a single cultivated female clone of this gynodioecious species (see p. 52), with information on the stamens added from Bitter (1910). The number of teeth is below the minimum given by Bitter (1910) for all but one of the seven subspecies which he recognizes, and this few-toothed form (subsp. *longiaristata* (Ross) Bitter var. *basipilosa* Bitter) is a small-leaved one, unlike ours.

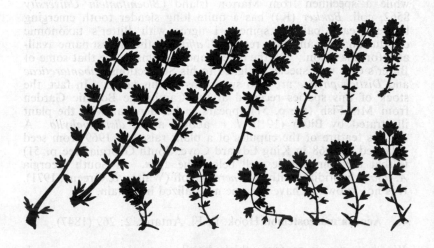

FIGURE 5
Above: *A.* 'Blue Haze' (207–56 Kew), × ½. Below: *A. glabra* (157A–71 Kew), × ½.

5. **Acaena 'BLUE HAZE'** C. R. Lancaster, New Cultivars of 1967, named by Hillier and Sons during 1967: 1 (1968).

Acaena adscendens hort., p.p.

Woody stems up to 4·5 mm thick; *herbaceous stems* coppery red, glabrous or very sparsely long-pilose or shortly glandular-pilose; sterile axillary shoots not rosetted. *Leaves* 1·8–7(–10)cm long; adnate portion of *stipule* 3–6mm long, foliaceous portion shorter or longer, entire or 2–6-lobed; rachis normally without subsidiary leaflets; *leaflets* (9–)11–13(–15), upper surface glaucous grey or grey-blue, more or less tinged purple, red or brown at the margins, with impressed venation, finely ciliate at the margins, lower surface more

strongly glaucous, veins often red, the midrib sometimes finely
pilose; the two *distal pairs of leaflets* more or less flabelliform or
broadly oblong, slightly longer than broad, or occasionally slightly
broader than long, 4·5–11mm long, 3·5–9·5mm wide, proximal margin
not decurrent, the *teeth* 5–9, obtuse, not penicillate, clefts extending
$\frac{2}{5}-\frac{1}{2}(-\frac{3}{5})$ of the way towards the midrib.   *Scapes* erect or ascending,
brownish red or deep red, 13–19cm long in fruit, finely appressed-
pilose; *capitula* spherical or very slightly longer than broad, appearing
brownish red, 7–10mm diam. in flower, 15–18mm in fruit (including
spines); *stamens* 2, filaments white, anthers red; *stigma* one, blackish
red, about as long as broad; *ripe cupules* c. 3·5mm long, 4-ribbed,
ribs bladdery above, more or less pilose distally; *spines* 4, 3–4mm
long, pinkish red, dilated at the base, barbed.   Fig. 5.

Origin unknown.

A specimen of this plant cultivated at Kew as *A. adscendens* in
1945 is in the Kew Herbarium, tentatively identified as *A. glabra* ×
A. anserinifolia.   To me it appeared more likely to be *A. glabra* ×
A. saccaticupula (both New Zealand species) as I could not find any
taxon ('pure' or hybrid) in Allan (1961) or Bitter (1910–1911) to
agree with it.   Lord Talbot de Malahide sent specimens to Dr
Lucy B. Moore (who worked on Allan's Flora both before and after
Allan's death) and she did not consider that it agreed with any of
the New Zealand species.   Mr Lancaster, in a document distributed
by Hillier and Sons of Winchester (see above) drew attention to
resemblances to *A. saccaticupula* and possibly to *A. microphylla*.
In 1969 I found that the plant comes true from seed, a fact which Lord
Talbot confirmed from his experience (by letter, August 1970).   In

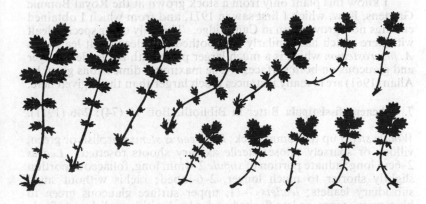

FIGURE 6

A. fissistipula (133–66 Copenhagen), ×½. Above: from greener-leaved plant with
more numerous stipule lobes. Below: from purple-leaved plant with fewer
stipule lobes.

view of this the plant must either be a hybrid which has become
stabilized (for example, by polyploidy or apomixis), or it must be a
naturally occurring form whose origin has been lost track of. In
the latter case the plant is probably South American, for it seems to
be closely similar in technical characters to *A. magellanica* and
A. affinis.

6.   **Acaena glabra** Buchanan in Trans. New Zeal. Inst. 4: 226 & t. 14
(1872).

Woody stems up to c. 3mm thick; *herbaceous stems* soon rooting,
greenish straw-coloured, glabrous; sterile axillary shoots elongate.
Leaves 3–5cm long; adnate portion of *stipule* 4–6mm long, foliaceous
portion slightly longer, usually 3-lobed; rachis without small subsidiary leaflets; *leaflets* (7–)9–11, rather thick and firm, upper surface
yellowish green, tinged brownish at edges, slightly glossy, smooth,
glabrous, lower surface glaucous grey-green, glabrous; the two
distal pairs of leaflets broadly cuneate-obovate, 1⅛–1½ times as long
as broad, 5·5–11mm long, 5–6·5mm wide, more or less petiolulate,
the *teeth* 5–8, obtuse, not penicillate, clefts extending ½-way to the
midrib or slightly more. *Scapes* erect, brownish, 6–10cm long in
fruit, glabrous; *capitula* spherical, 6–7·5mm diam. in flower, c. 11mm
in fruit (including spines); *stamens* 2, white; *stigma* one, c. 3 times
as long as broad; *ripe cupules* c. 2·5mm long, with 4 bladdery lobes
at the top, 2 of them expanded to form wings, glabrous; *spines* 4,
up to 2mm long but perhaps sometimes vestigial, not barbed. Fig. 5.

Native in New Zealand (Southern Alps).

I know this plant only from a stock grown at the Royal Botanic
Gardens, Kew, which I first saw in 1971, and from which I obtained
cuttings now growing on in Cambridge. The only other species dealt
with here which has similarly few-toothed and deeply cut leaflets is
A. macrostemon which is a much larger plant with the leaflets longer
and glaucous on both surfaces. The maximum dimensions given by
Allan (1961) are in many instances much larger than those given here.

7.   **Acaena fissistipula** Bitter in Biblioth. Bot. 17 (74): 246 (1911).

Woody stems up to 2·5mm thick; *herbaceous stems* purplish or green,
villous or sparsely pilose; sterile axillary shoots rosetted. *Leaves*
2–6cm long; adnate portion of *stipule* 2–4mm long, foliaceous portion
slightly shorter to much longer, 2–6-lobed; rachis without small
subsidiary leaflets; *leaflets* 7–11, upper surface glaucous green to
blue-green, more or less flushed and edged with purplish red, finely
appressed-pilose or glabrous, lower surface paler, pilose; the two
distal pairs of leaflets broadly oblong to suborbicular, 1–1⅓ times as
long as broad, 4–10mm long, 3–8mm wide, margin not decurrent,
the *teeth* 6–9, obtuse to acute, slightly penicillate, clefts extending

FIGURE 7
A. ovalifolia (106–69 McClintock), ×1.

$\frac{1}{4}$–$\frac{1}{3}$ of the way towards the midrib. *Scapes* erect or ascending, greenish or brownish, pilose to villous, 6–13cm long in fruit; *capitula* spherical, 4·5–7·5mm diam. in flower, 12–13mm in fruit (including spines); *stamens* 2, filaments white, anthers red or flushed rose; *stigma* one, deep red, 2–3 times as long as broad; *ripe cupules* 1·5–2·5mm long, 4-angled, silky-hairy; *spines* 4, 2–4mm long, reddish, barbed. Fig. 6.

Native in New Zealand (mountains of South Island).

The above description is based on two stocks growing at Cambridge, one from Christchurch Botanic Garden, New Zealand, and one from the University Botanic Garden, Copenhagen. The former is purplish-tinged but the latter is a mixture of purplish-tinged and glaucous green growth, presumably representing different plants growing intertwined. The greener growth (Fig. 6, upper) has the leaflets relatively narrower than the purple (Fig. 6, lower) and the stipules with more numerous lobes; possibly it represents a hybrid, but there is no clear character to separate it from *A. fissistipula*.

8.   **Acaena ovalifolia** Ruiz & Pavón, Fl. Peruv. 1: 67 & t. 103c (1798).

Woody stems up to c. 3mm thick; *herbaceous stems* green, sometimes flushed red, pubescent with long fine white hairs; axillary shoots elongate. *Leaves* 5–12cm long; adnate portion of *stipule* 3–5mm long, foliaceous portion longer, 2–4-lobed; rachis without small subsidiary leaflets; *leaflets* (7–)9, upper surface bright green and rugulose, sometimes glossy when young, glabrous, lower surface glaucescent, pilose, silky when young, especially on the midrib and main veins; the two *distal pairs of leaflets* elliptic or oblong, 1$\frac{3}{4}$–2 times as long as broad, (10–)15–30mm long, (5–)9–16mm wide, margins not decurrent, the *teeth* (12–)17–23, subacute or acute, not penicillate, clefts extending $\frac{1}{4}$–$\frac{1}{3}$ of the way towards the midrib. *Scapes* ascending, green densely appressed-pilose or somewhat spreadingly pilose, (2–)6–12cm long in fruit; *capitula* spherical, 8–10mm diam. in flower, 18–30mm in fruit (including spines); *stamens* 2, white; *stigma* one, white, slightly longer than broad; *ripe cupules* c. 3mm long, pedicelled, bladdery at the apex, silky-hairy; *spines* 2, 8–10mm long, red, barbed. Fig. 7.

Native in South America from the Magellan region northwards through the cordillera of the Andes to Colombia.

Introduced. V.C. 1: Great Erth, 1951, *J. E. Lousley* (BM). V. C. 3: N. of Ivybridge, 1968 (cult. 1970, 1971), *M. A. Turner* (CGG). V. C. 82, W. of Gullane, near Dirleton, 1969 (cult. 1972), *D. McClintock* (CGG). V. C. 96: Blackfold, roadside, 1961, *M. McCallum Webster* 5830 (CGE), & 1962, 7527 (BM, CGE). V.C. 98: on a wall top at Ardtornish, 1971, *A. M. Hugh Smith* (CGE);

FIGURE 8
A. novae-zelandiae (192B–68 Dawson), ×½.

edge of woodland, Poltalloch, N. of Lochgilphead, 1968, *D. McClintock* (CGE). V.C. 99: near waterworks above Rhu, 1966, *J. E. Lousley* (BM). V.C. Hl: Dromquinna Estate, Kenmare, (cult. Bayfordbury), n.d. (?c. 1960), *K. Beckett* (CGE). V.C. H13: woods of Fenagh House, national grid 03/62, 1962, *E. Booth* (*M. McCallum Webster* 8002) (BM).

This species resembles *A. novae-zelandiae* in its bright green oblong or elliptic leaflets and in the long slender red spines of the cupule, but it differs in being generally a larger plant and in the scarcely glossy leaflets with many teeth.

Four subspecies of *A. ovalifolia* are recognized by Bitter (1910).

9.   **Acaena novae-zelandiae** Kirk in Trans. New Zeal. Inst. 3: 177 (1871).

Acaena sanguisorbae (L. f.) Vahl subsp. *novae-zelandiae* (Kirk) Bitter in Biblioth. Bot. 17 (74): 263 (1911).
Acaena anserinifolia auctt. Brit. et Austral.
Acaena sanguisorbae auctt. Austral.

Woody stems up to 5mm thick; *herbaceous stems* pale green, more or less deeply flushed pinkish red, sparsely to densely villous; sterile axillary shoots rosetted.   *Leaves* 3–6(–10)cm long; adnate portion of *stipule* 2–5mm long, foliaceous portion longer, entire or up to 4-lobed; rachis without small subsidiary leaflets; *leaflets* 9–13(–15), upper surface bright green, glossy but rugulose, glabrous or with a few appressed hairs, lower surface glaucous green, sparsely to densely silky-hairy; the two *distal pairs of leaflets* oblong or obovate-lanceolate, just less than twice, or up to 2½ times, as long as broad, (4·5–)7–16(–19)mm long, (2–)3–6(–10)mm wide, proximal margin sometimes slightly decurrent, the *teeth* 8–15, acute to obtusish, not penicillate, clefts extending c. ⅓ of the way towards the midrib. *Scapes* pale green or flushed reddish, appressed-pilose, 4–11cm long in fruit; *capitula* spherical, 6–9mm diam. in flower, 15–30mm in fruit (including spines); *stamens* 2, white; *stigma* one, white, tipped purplish, c. 3 times as long as broad; *ripe cupules* 3·5–4mm long, 4-ribbed, villous; *spines* 4, 6–9·5mm long, red, barbed.   Fig. 8.

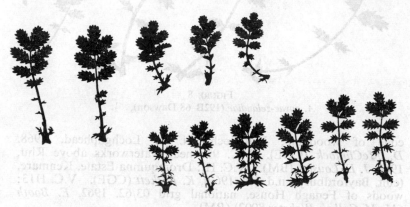

FIGURE 9

A. anserinifolia, × ½. Above: 192A–68 Dawson. Below: 208–65 Anderson.

Native in New Zealand (North and South Islands, lowland to lower montane), Australia, Tasmania.

Introduced. V.C. 3: Yarner Wood, Bovey Tracey, 1957, *S. M. Walters* (CGE) & (cult. 1970), *D. W. H. Walton* (CGG); Dartmoor, 1943, *J. E. Raven* (CGE); roadside near Newton Abbot, 1951, *M. Brown* (CGE).   V.C. 9: near Studland, 1938, *W. Nielson-Jones* (BM); Ballard Down, Studland, 1953, *M. F. Hancock* (BM). V.C. 12: Blackmoor, 1964 (cult. 1965), *J. E. Lousley* W/2623 (BM); Blackmoor, 1965 (cult. 1965), *M. McCallum Webster* 10193 (BM). V.C. 16: Mereworth Wood, 1947, 1949, *J. E. Lousley* (BM). V.C. 17: Witley Common, where established for c. 20 years in two places, 1959, *J. E. Lousley* (BM).   V.C. 26: Wangford Warren, 1964, *O. Rackham* (CGE).   V.C. 27: Kelling Heath between Holt and Weybourne, 1935, *C. L. Collenette* (BM); 1964, *A. S. Watt* (CGE); 1952, *V. Maynard* (BM); 1967, *P. F. Yeo* 671 (CGE), & (cult. 1969) (CGG).   V.C. 37: Severn Stoke, 1966, *J. E. Lousley* & *J. Russell* (BM).   V.C. 64: Meanwood, Leeds, naturalized in grounds of Alder Hill, 1914, *F. A. Lees* (BM, mixed with *A. anserinifolia*), & 1917 (BM).   V.C. 68: Holy Island, 1955, *B. Sowerby* 185 (BM) & 1955, *J. E. Lousley* (BM).   V.C. 80: junction of rivers Tweed and

Gala, 1916, *G. C. Druce* (BM); banks of the river Tweed at Melrose, 1912, *I. M. Hayward* (BM, CGE). V.C. 81: grounds of estate at Paxton, Hutton parish, 1960, *E. M. C. Isherwood & W. Frost* (BM).

This is the species described by Warburg (in Clapham, Tutin & Warburg (1962)) and by Valentine in *Flora Europaea* (Tutin *et al.*, (1968)) as *A. anserinifolia*. Of the above records those from vice-counties 12, 37 and 64 are stated to have been introduced with wool shoddy.

Acaena novae-zelandiae occurs not only in New Zealand, but also in the mountainous parts of Australia and Tasmania; a larger and coarser form than the typical one, with more obovate leaflets, occurs mainly by the sea in the south of the continent, and there is one specimen of this form from Tasmania in the Kew Herbarium. Seed which was kindly sent to me by Dr Hj. Eichler from Messent National Park, South Australia, in 1968 (*Eichler* 19725, Dec. 1967), of which plants have been grown at Cambridge, appears to be of this form (CGG). The plants (not covered by the description above) have the hair-bases of the stem pustulate, the flowering shoots erect, standing 15–25cm high (10–15cm in New Zealand forms growing alongside), the leaves usually 4–10cm long (commonly more than 6cm), and the distal leaflets 9–22mm long (commonly more than 15mm) and correspondingly wide*. The plants have survived out-of-doors since 1968 but appear to be cut back more severely by winter frost than New Zealand forms. A plant received from Mr McClintock in 1968 represents this form; it was growing in the grounds of Pantycelin Hall, University College, Aberystwyth (CGG).

The New Zealand form of *A. novae-zelandiae* has been grown at Cambridge from seed supplied by Christchurch Botanic Garden and by Dr J. W. Dawson, who collected it at Crofton Downs, Wellington. Plants from various garden sources and that from Yarner Wood (see above) agree with the New Zealand form in their characters. Plants from Kelling Heath, Norfolk, have been grown at Cambridge, and have rather broader leaflets (distal pairs $1\frac{1}{3}$–$1\frac{2}{3}$(–2) times as long as broad). The cupular spines on my plants are not more than 4mm long but in Dr Watt's specimens they are up to 8mm long. In New Zealand this species hybridizes with *A. anserinifolia*; the hybrids are uniform and probably of the F_1 generation, but segregating progeny were raised from them in cultivation (Dawson, 1960). The possibility exists that the Kelling Heath population, which is uniform, is a more or less stabilized derivative of such a cross. However, it scores only 5 in Dawson's hybrid index out of a maximum of 25 for *A. anserinifolia*, which is little more than the usual 0–1 for *A. novae-zelandiae*. It can therefore safely be treated as falling within the latter species.

---

*They lack the large cotyledons and long spines of *A. pallida* (Kirk) Allan, a similar large coastal form from New Zeland (Dawson, 1960).

10. **Acaena anserinifolia** (J. R. & G. Forster) Druce in Rep. Bot.
Exch. Club Brit. Is. 1916: 484 (1917).

Ancistrum anserinifolium J. R. & G. Forster, Char. Gen. Pl. 4
(1776).
Ancistrum sanguisorbae L. f., Suppl., 89 (1781), nomen superfl.
Acaena sanguisorbae (L.f.) Vahl subsp. *profundeincisa* Bitter in
Biblioth. Bot. 17 (74): 270 (1911).

Woody stems up to c. 3mm thick; *herbaceous stems* brown, sparsely
to densely silky-pubescent; sterile axillary shoots rosetted. *Leaves*
2–5cm long; adnate portion of *stipule* 2–3mm long, foliaceous portion
longer, 3–8-lobed; rachis without small subsidiary leaflets; *leaflets*
9–13, upper surface light matt green, more or less flushed brown, the
basal sometimes entirely brown, sparsely appressed-silky, lower
surface glaucescent green, sometimes purplish-tinged, sparsely to
densely appressed-silky; the two *distal pairs of leaflets* oblong or
oblong-obovate, ($1\frac{1}{4}$–)$1\frac{1}{3}$–2 times as long as broad, 3–8(–10)mm long,
1·5–4·5(–6)mm wide, proximal margin of one or both pairs decurrent
for 0·5–1mm, the *teeth* (6–)10–12(–13), acute or subacute, penicillate,
clefts extending $\frac{1}{3}$(–$\frac{1}{2}$) of the way towards the midrib. *Scapes* erect
or ascending, brown, appressed-silky, 3–7·5(–9)cm long in fruit;
capitula spherical, 5–8mm diam. in flower, 12–16mm in fruit
(including spines); *stamens* 2, white; *stigma* one, white, about as
long as broad; *ripe cupules* 2–2·5mm long, 4-ribbed, silky-hairy;
spines 4, 3·5–4·5(–6)mm long, red, barbed.   Fig. 9.

Native in New Zealand (North and South Islands, lowland to
montane), Australia(?).
Introduced. V.C. 64: Alder Hill, Meanwood, Leeds (brought
into gardens with wool shoddy), 1914, *F. A. Lees* (BM, mixed with
A. novae-zelandiae). V.C. 88: St. Fillans, Loch Earn, 1970,
J. M. Naylor per *D. McClintock* (herb D. McClintock).

As explained under *A. novae-zelandiae,* most British material
named *A. anserinifolia* is the former species. More difficult to
separate is *A. pusilla* (no. 11).
The hybrid *A. anserinifolia* × *A. inermis* is reported to occur in
New Zealand (Allan, 1961: 363), and is in cultivation at Cambridge.
Two extremely similar plants were received in 1969 (they possibly
represent the same clone, though at times slight differences are
apparent), one from Batworthy Brook, Dartmoor, S. Devon, collected by Mr R. J. Skerrett, and one from the garden of Mr Michael
Noble, Strone, Cairndow, Argyll, sent by Mr McClintock. These
plants have relatively broad leaflets (Fig. 12) of a greyish olive-green
colour at flowering time, but even so could be taken for a form of
A. anserinifolia since they have silky-looking foliage and spiny
cupules. However, the stolons root very rapidly and in most flowers
there are two styles (as in *A. inermis*), only a few flowers having a

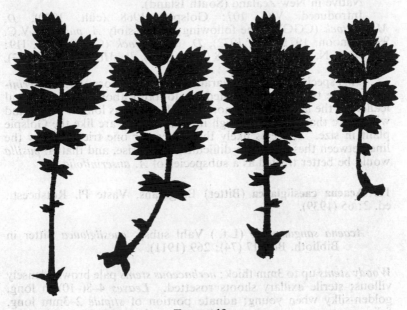

FIGURE 10
A. caesiiglauca (103–60 Kew), ×1.

single style.   Mr McClintock sent seed with his plant and from these, seedlings were raised, six of which were grown to maturity.   These plants show remarkable variation in colour, three having leaves coloured rather like *A. anserinifolia* (though in one the brown is paler and in one redder), one is dark purplish grey-green, one pure blue-green glaucous and one coppery greenish brown; one plant has two stigmas and four have one (one plant was lost before the flowers were examined).

11. **Acaena pusilla** (Bitter) Allan, Fl. New Zeal. 1: 362 (1961).

Acaena sanguisorbae (L.f.) Vahl subsp. *pusilla* Bitter in Biblioth. Bot. 17 (74): 271 (1911).

Very similar to *A. anserinifolia* but with *woody stems* only up to 2mm thick, leaves 1–2.5 cm long, *stipule* with adnate portion not more than 2mm long, foliaceous portion with 2–5-lobes, *leaflets* 7–9, the two *distal pairs* not more than 1½ times as long as broad, not more than 4·5(–6)mm long and 3·5(–4)mm wide with 5–9 teeth, *scapes* in fruit 2–5cm long, *capitula* in flower 4–5mm diam., in fruit up to 13mm (including spines), *ripe cupules* c. 1·5mm long, spines not more than 4(–6)mm long.   Fig. 11.

Native in New Zealand (South Island).
Introduced. V.C. 107: Golspie, 1968 (cult. 1970), *D. McClintock* (CGG). The following are possibly *A. pusilla*:—V.C. 99: Dunoon, 1971 (cult. 1972), *D. McClintock* (CGG); V.C. H9: Aasleagh, N. of Leenane, 1968 (cult. 1970), *D. McClintock* (CGG).

This species has the appearance of a dwarf form of *A. anserinifolia*. The two gatherings cited as doubtful above have the distal leaflets of the largest leaves more than 1½ times as long as broad and with more than 9 teeth, though in general they are like the Golspie plant in size. It seems likely that wherever one tries to draw the line between these two taxa difficulties will arise, and that *A. pusilla* would be better treated as a subspecies of *A. anserinifolia*.

12. Acaena caesiiglauca (Bitter) Bergmans, Vaste Pl. Rotsheest., ed. 2: 65 (1939).

Acaena sanguisorbae (L.f.) Vahl subsp. *caesiiglauca* Bitter in Biblioth. Bot. 17 (74): 269 (1911).

Woody stems up to 3mm thick; *herbaceous stems* pale brown, densely villous; sterile axillary shoots rosetted. *Leaves* 4–8(–10)cm long, golden-silky when young; adnate portion of *stipule* 2–3mm long, foliaceous portion as long or longer, entire or 2–3-lobed; rachis

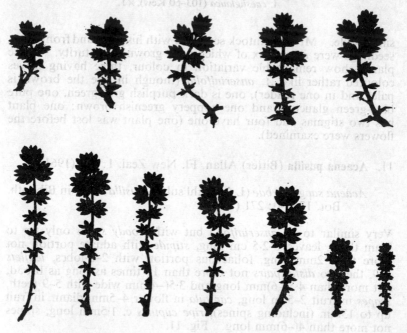

FIGURE 11

Above: *A. pusilla* (200C–68 McClintock), ×1·4. Below: *A. microphylla* (207–56. Kew), ×1·4.

without small subsidary leaflets; *leaflets* 7–9, upper surface blue-grey glaucous, becoming lightly purple-tinged in age, sparsely appressed-pilose, lower surface slightly paler and more glaucous, becoming purple-tinged in age, silky-hairy; the two *distal pairs of leaflets* oblong or broadly obovate, $1\frac{1}{3}$–$1\frac{1}{2}$ times as long as broad, 6–14mm long, 5–10mm wide, margins not decurrent, the *teeth* 6–10, acute, penicillate, clefts extending $\frac{1}{4}$–$\frac{1}{3}$ of the way towards the midrib. *Scapes* erect, pale brown, villous, 10–14cm long in fruit; *capitula* spherical, 7–11mm diam. in flower, up to 23mm in fruit (including spines); *stamens* 2, white; *stigma* one, white or flushed pink, c. 4 times as long as broad; *ripe cupules* 3·5mm long, 4-ribbed, densely hirsute; *spines* 4, 4·5–7mm long, olive-brown, barbed. Fig. 10.

Native in New Zealand (South Island, montane and subalpine).

This is perhaps the most beautiful of all the Acaenas, especially when it acquires its purple tinge; it is known in the British Isles only as a garden plant. Allan (1961) described a var. *pilosa* (Kirk) Allan having 11–13 leaflets with the upper surface densely hirsute.

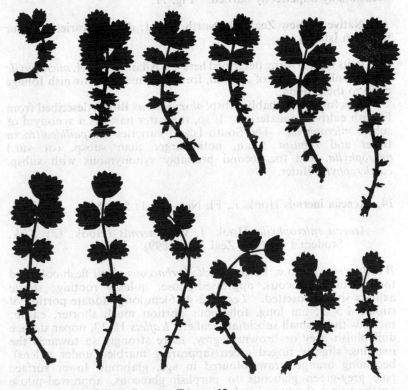

FIGURE 12

Above: *A. anserinifolia* × *A. inermis* (166A–69 McClintock), × ·1. Below: *A. inermis* (38–69 Glasnevin), × ·1.

13. **Acaena microphylla** Hook. f., Fl. Nov.-Zel. 1: 555 (1852).

Woody stems up to c. 2mm thick; *herbaceous stems* light brown or brownish green, appressed-pilose or glabrous, quickly rooting; axillary shoots rosetted. *Leaves* 1–3cm long; adnate portion of *stipule* 1–2mm long, foliaceous portion as long or shorter, entire; rachis without small subsidiary leaflets; *leaflets* 11–13, upper surface yellowish green, strongly flushed brownish towards the edges, becoming entirely dull purplish brown, glabrous, lower surface similarly coloured but glaucescent, appressed-pilose on the veins; the two *distal pairs of leaflets* square or oblate, as long as broad or slightly shorter, (1·5–)2–4·5mm long, 1·5–4·5mm wide, margins not decurrent, the *teeth* 3–7, obtuse, penicillate, clefts extending $\frac{1}{3}$–$\frac{3}{5}$ of the way towards the midrib. *Scapes* ascending, brown, pilose, 1–4cm long in fruit; *capitula* spherical, 5–6mm diam. in flower, up to 30mm in fruit (including spines); *stamens* 2, white; *stigmas* 2, white with pink rachis, 2–3 times as long as broad; *ripe* cupules c. 1·75mm long, broadly turbinate, 4-angled, glabrous, reddish; *spines* 4, up to 13mm long, soft, thick, bright red, not barbed, or occasionally imperfectly barbed. Fig. 11.

Native in New Zealand (North Island; distinct varieties occur in South Island).

In its typical form, described here (cf. Allan, 1961), *A. microphylla* is the smallest species of *Acaena,* forming a mat of brownish foliage close to the ground.

This form is probably subsp. *obscurascens* Bitter, described from English cultivated material. If so, the latter name is a synonym of subsp. *microphylla.* The South Island varieties are *pallideolivacea* Bitter and *robusta* Allan, both larger than subsp. (or var.) *microphylla,* and the second probably synonymous with subsp. *eumicrophylla* Bitter.

14. **Acaena inermis** Hook. f., Fl. Nov-Zel. 1: 54 (1852).

Acaena microphylla Hook. f. var. *inermis* (Hook. f.) Kirk, Student's Fl. New Zeal. 134 (1899).

Woody stems up to c. 1·5mm thick; *herbaceous stems* flesh-coloured to brownish glaucous, appressed-pilose, quickly rooting; sterile axillary shoots rosetted. *Leaves* 2–4(–6)cm long; adnate portion of stipule 1·5–2·5mm long, foliaceous portion much shorter, entire; rachis without small subsidiary leaflets; *leaflets* 11–13, upper surface dull bluish grey or brownish grey, more strongly so towards the margins, slightly tinged green (appearing marbled under a lens), becoming orange straw-coloured in age, glabrous, lower surface pale grey-green glaucous to purplish glaucous, appressed-pilose on the veins; the two *distal pairs of leaflets* flabelliform, square or oblate, as long as broad or shorter, 2–5(–8)mm long and wide, margins not decurrent, the teeth 5–10, obtuse, penicillate, clefts extending ¼–⅖ of the way towards the midrib. *Scapes* erect, brownish, villous or appressed pilose, 1–4(–6)cm long in fruit; *capitula* spherical, 5–6·5mm diam. in flower, 7mm in fruit; *stamens* 2(?–4), white; *stigmas* 2, white, c. twice as long as broad; *ripe cupules* c. 1·5mm long, broadly turbinate, with 4 bladdery lobes distally, pubescent; *spines* none. Fig. 12.

New Zealand (South Island, montane).
Introduced. V.C. 103: driveway to Glengorm Castle, 1968, *Brit. Mus. Mull Survey* 2554 (BM), & 1969, 4019, 4020 & 4021 (cult. Cambridge–239–69 *Ferguson*–1970) (CGG). V.C. 104: Raasay, 1968 (cult. Cambridge, 1970, 1972) *P. Boyd* per *C. H. Gimingham* & *D. McClintock* (CGG), & 1969 (CGE—see below).

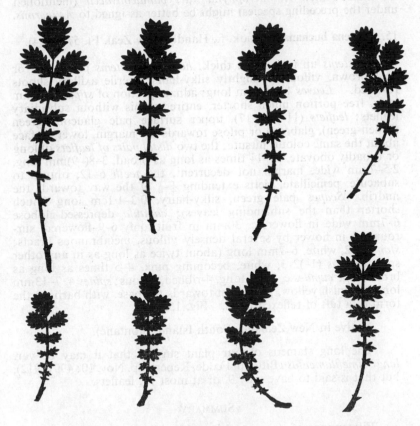

FIGURE 13
A. buchananii (1–56 Chelsea), ×1·1.

This dwarf, closely prostrate species is grown in gardens and has escaped from them; the Mull record is an example and Mr McClintock has sent me specimens from Pantycelin Hall, University College of Aberystwyth, where it maintains itself and spreads. I have also had this species from Glasnevin Botanic Garden and from Mr David Walton. The plant from Raasay was first determined as *A. microphylla* because the wild specimen has cupules with the spines of that species. In cultivation I find that the spines develop rather late, and on some cupules not at all. Otherwise the living plants look exactly like my other cultivated stocks (except for the Glengorm Castle one which is more robust and has more purplish leaves). Spined and spineless forms were also noticed in the Glengorm Castle material by Mr McClintock. The possibility of hybridity cannot be ruled out, but it seems that forms producing spined cupules may have to be accepted within *A. inermis,* and that the short-spined *A. microphylla* var. *pallideolivacea* (mentioned under the preceding species) might be better assigned to *A. inermis.*

15. **Acaena buchananii** Hook. f., Handb. New Zeal. Fl. 57 (1864).

Woody stems up to c. 2mm thick; *herbaceous stems* pale green or pale brown, villous or slightly silky-hairy; sterile axillary shoots rosetted.   *Leaves* 1·5–5·5cm long; adnate portion of *stipule* 3–8mm long, free portion much shorter, entire; rachis without subsidiary leaflets; *leaflets* (11–)15(–17), upper surface pale glaucous green (lichen-green), glabrous or pilose towards the margin, lower surface about the same colour, hirsute; the two *distal pairs of leaflets* oblong or broadly obovate, $1\frac{1}{4}$–$1\frac{1}{2}$ times as long as broad, 3–8(–9)mm long, 2·5–7mm wide, margin not decurrent, the *teeth* 6–12, obtuse to subacute, penicillate, clefts extending $\frac{1}{4}$–$\frac{1}{3}$ of the way towards the midrib.   *Scapes* pale green, silky-hairy, 0·3–1·1cm long (much shorter than the subtending leaves); *capitula* depressed-globose 6–7mm wide in flower, c. 30mm in fruit, only 6–9–flowered, surrounded in flower by several densely villous, membranous bracts; *stamens* 2, white, 6–7mm long (about twice as long as in any other sp.); *stigmas* (1–)2–3, white, becoming pink, 4–5 times as long as broad; *ripe cupules* c. 2mm long, 4-ribbed, villous; *spines* 4, 7–13mm long, greenish yellow, pubescent towards the base, with barbs in the form of a tuft of reflexed hairs.   Fig. 13.

Native in New Zealand (South Island, montane).

The long stamens of our plant suggest that it may be var. *longissimefilamentosa* Bitter in Fedde, Repert. Sp. Nov. 10: 499 (1912), but that is said to have only 9, or at most 11, leaflets.

SUMMARY

Fifteen taxa of *Acaena* cultivated and naturalized in the British Isles are described and annotated. A vegetative key is provided.

A. anserinifolia is rarely naturalized, most material under this name being *A. novae-zelandiae* (which in Australia is also known as *A. anserinifolia*). The only naturalized member of the *A. magellanica* group is *A. magellanica* subsp. *magellanica*, material of which has been referred to *A. laevigata* (*A. magellanica* subsp. *laevigata*) or *A. adscendens*. The plant to which the name *A. adscendens* is usually applied in botanical literature should be called *A. affinis*. Garden material named *A. adscendens* is usually *A. affinis* or a true-breeding plant of unknown origin for which the cultivar name *A.* 'Blue Haze' is available.

ACKNOWLEDGMENTS

I wish to thank all those who kindly gathered specimens, especially David McClintock, who has been most assiduous, and the authorities in charge of the herbaria at Kew and the British Museum (Natural History) for facilities for study. To my thanks to David Walton for specimens donated I have to add appreciation of his kindness in joining in lengthy discussions on visits to Cambridge and letting me see numerous specimens loaned to him by foreign herbaria.

REFERENCES

ALLAN, H. H. (1961). *Flora of New Zealand*, 1.
BITTER, G. (1910–1911). Die Gattung *Acaena*. *Biblioth. Bot.* **17** (74): 1–248 (1910) & 249–336 (1911).
—— (1912). Weitere Untersuchungen über die Gattung Acaena. *Fedde, Repert. Sp. Nov.* **10**: 489–501.
DAWSON, J. W. (1960). Natural *Acaena* hybrids in the vicinity of Wellington. *Trans. Roy. Soc. New Zeal.* **88**: 13–27.
GRONDONA, E. (1964). Las especies argentinas del género "Acaena" ("Rosaceae"). *Darwiniana* **13**: 208–342.
MOORE, D. M. (1968). *The Vascular Flora of the Falkland Islands.* (British Antarctic Survey, Scientific Report no. 60).
VALENTINE, D. H. (1968). In Tutin, T. G., *et al.*, eds., *Flora Europaea* **2**.
WALTON, D. W. H. & GREENE, S. W. (1971). The South Georgian species of *Acaena* and their probable hybrid. *Brit. Antarct. Surv. Bull.* **25**: 29–44.
WARBURG, E. F. (1962). In Clapham, A. R., Tutin, T. G. & Warburg, E. F., *Flora of the British Isles*, ed. 2.

INDEX

Where a plant is referred to in the text by its English name, reference is included in the index both under this name and its scientific name; where no English name is used in the text the index lists the scientific name only. The numerous cultivars of British trees and shrubs exhibited by Messrs Hilliers & Sons are not listed individually, they may be traced from the index via the name of their species.

In Acaena, *page numbers in bold print refer to the main taxonomic account of the individual taxon in Appendix III.*